Letts

Revise AS

ED

Edexcel Biology

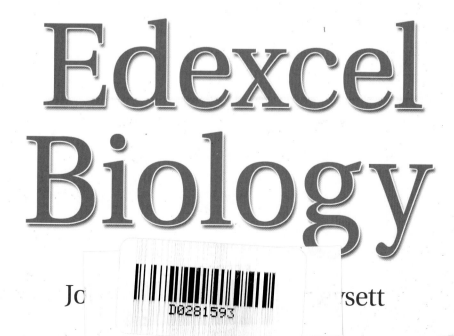

Jo   rsett

Contents

Chapter 3 Enzymes

Chapter 4 Exchange

Chapter 5 Transport

Contents

Specification list

The specification labels on each page refer directly to the units in the exam specification, i.e. EDEXCEL ⟩ 4.2 refers to unit 4, section 2.

Edexcel Biology

MODULE	SPECIFICATION TOPIC	CHAPTER REFERENCE	STUDIED IN CLASS	REVISED	PRACTICE QUESTIONS
Unit 1	Biological molecules	1.1, 1.2, 1.3, 1.4			
	Cell membranes and transport	4.1, 4.2, 4.3			
	Transport in animals	5.2, 5.3, 8.1			
	DNA and protein synthesis	6.1			
	Inheritance	8.2			
	Mutations	7.2			
Unit 2B	Cell structure	2.1, 2.2			
	Cell division	6.2			
	Variation	7.2			
	Biodiversity	7.2			
	Classification	7.1			

Examination analysis

Unit 1 AS
This unit is assessed by means of a written examination paper, which lasts 1 hour 15 minutes and will include: objective questions, structured questions, short-answer questions and will also cover How Science Works and practical-related questions.   80 marks

Unit 2 AS
This unit is assessed by means of a written examination paper, which lasts 1 hour 15 minutes and will include: objective questions, structured questions, short-answer questions and will also cover How Science Works and practical-related questions.   80 marks

Unit 3 AS
Students will submit a written report of between 1500 and 2000 words which will be marked by the teacher and moderated by Edexcel. The report may be either a record of a visit to a site of biological interest or a report of research into a biological topic. During the course teachers will observe students carrying out practical work and submit their assessment (non-moderated). There is no separate content for this unit.   50 marks

The AS/A2 Level Biology course

AS and A2

The Edexcel Biology A Level course being studied from September 2008 is in two parts, with a number of separate modules or units in each part. Most students will start by studying the AS (Advanced Subsidiary) course. Some will go on to study the second part of the A Level course, called A2. It is also possible to study the full A Level course in either order. Advanced Subsidiary is assessed at the standard expected halfway through an A Level course, i.e. between GCSE and A Level. This means that the new AS and A2 courses are designed so that difficulty steadily increases:

- AS Biology builds from GCSE Science and Additional Science/Biology.
- A2 Biology builds from AS Biology.

How will you be tested?

Assessment units

AS Biology comprises three units or modules. The first two units are assessed by examinations. The third component involves assessment of practical and research skills.

The practical skills are marked by your teacher. The research project is marked by your teacher and the marks can be adjusted by moderators appointed by Edexcel.

For the full A Level in Biology, you will take a further three units. AS Biology forms 50% of the assessment weighting for the full A Level.

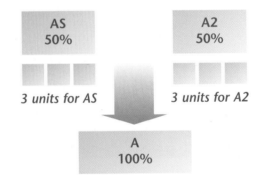

AS 50%				A2 50%		
3 units for AS				*3 units for A2*		

A
100%

Tests are taken at two specific times of the year, January and June. It can be an advantage to you to take a unit test at the earlier optional time because you can re-sit the test. The best mark will be credited and the lower mark ignored.

Each unit can normally be taken in either January or June. Alternatively, you can study the whole course before taking any of the unit tests. There is a lot of flexibility about when exams can be taken and the following diagram shows just some of the ways that the assessment units may be taken for AS and A Level Biology.

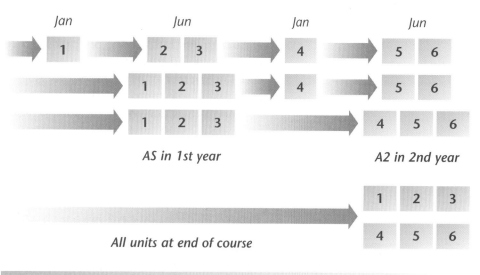

Jan	Jun	Jan	Jun

AS in 1st year   *A2 in 2nd year*

All units at end of course

A2 and synoptic assessment

Many students who have studied at AS Level may decide to go on to study A2. There are three further units or modules to be studied. The A Level specification includes a 'synoptic' assessment at the end of A2. Synoptic questions make use of concepts from earlier units, bringing them together in holistic contexts. Examiners will test your ability to inter-relate topics through the complete course from AS to A2.

Internal Assessment

Assessment of practical skills and a research project form part of your A Level Biology course.

What skills will I need?

For AS Biology, you will be tested by assessment objectives: these are the skills and abilities that you should have acquired by studying the course. The assessment objectives for AS Biology are shown below.

Knowledge with understanding

- recall of facts, terminology and relationships
- understanding of principles and concepts
- drawing on existing knowledge to show understanding of the responsible use of biological applications in society
- selecting, organising and presenting information clearly and logically

Application of knowledge and understanding, analysis and evaluation

- explaining and interpreting principles and concepts
- interpreting and translating, from one to another, data presented as continuous prose or in tables, diagrams and graphs
- carrying out relevant calculations
- applying knowledge and understanding to familiar and unfamiliar situations
- assessing the validity of biological information, experiments, inferences and statements

You must also present arguments and ideas clearly and logically, using specialist vocabulary where appropriate. Remember to balance your argument!

Practical Biology and research skills

Your practical skills will be assessed by your teacher throughout the course. They will look at your ability to:
- use apparatus skilfully and safely
- produce and record reliable and valid results
- present and analyse data.

It is therefore important in these tasks to know the difference between **validity** and **reliability**.

Validity: in a valid experiment you only investigate one variable at a time. All the other variables are kept constant. This means that you are really measuring what you are trying to measure.

Reliability: an experiment is reliable if it can be repeated and similar results are obtained. Reliability can usually be increased by taking repeat readings and averaging them.

For the assessment of your research skills you will have to produce a written report. This is based on a visit to a site of biological interest or a report of research into a biological topic. You are not expected to do any practical work.

The report must:
- contain between 1500 and 2000 words
- be word-processed
- have a clear structure with technical vocabulary
- include visual methods of presentation such as graphs and tables
- acknowledge any sources in a bibliography.

Different types of questions in AS examinations

In AS Biology examinations different types of questions are used to assess your abilities and skills. Unit tests mainly use structured questions requiring both short-answers and more extended answers.

Short-answer questions

A question will normally begin with a brief amount of stimulus material. This may be in the form of a diagram, data or graph. A short-answer question may begin by testing recall. Usually this is followed up by questions which test understanding. Often you will be required to analyse data.

Short-answer questions normally have a space for your responses on the printed paper. The number of lines is a guide as to the amount of words you will need to answer the question. The number of marks indicated on the right side of the paper shows the number of marks you can score for each question part.

Here are some examples. (The answers are shown in blue)

The diagram shows part of a DNA molecule.

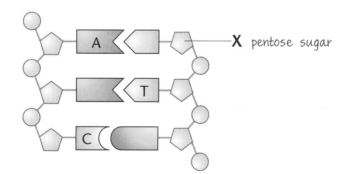

(a) Label part X.   [1]

(b) Complete the diagram by writing a letter for each missing organic base in each empty box.   [1]

(b) How do two strands of DNA join to each other?

 The organic bases ✓ link the strands by hydrogen bonds ✓   [2]

Structured questions

Structured questions are in several parts. The parts are usually about a common context and they often progress in difficulty as you work through each of the parts. They may start with simple recall, then test understanding of a familiar or unfamiliar situation. If the context seems unfamiliar the material will still be centred around concepts and skills from the Biology specification. (If a student can answer questions about unfamiliar situations then they display understanding rather than simple recall.)

The most difficult part of a structured question is usually at the end. Ascending in difficulty, a question allows a candidate to build in confidence. Right at the end technological and social applications of biological principles give a more demanding challenge. Most of the questions in this book are structured questions. This is the main type of question used in the assessment of AS Biology.

When answering structured questions, do not feel that you have to complete a question before starting the next. Answering a part that you are sure of will build your confidence. If you run out of ideas go on to the next question. This will be more profitable than staying with a very difficult question which slows down progress. Return at the end when you have more time.

Here is an example of a structured question which becomes progressively more demanding.

Question

The diagram shows the molecules of a cell surface membrane.

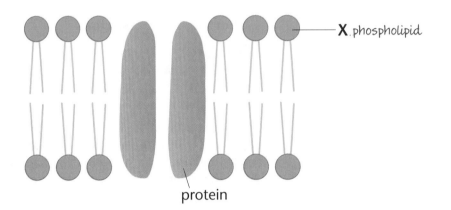

(a) (i)   Label molecule X.   [1]

   (ii) The part of molecule X facing the outside of a cell is hydrophilic.
         What does this mean?

         water loving/water attracting   [1]

   (iii) Describe **one** feature of the part of molecule X which faces inwards.

         Hydrophobic/ water hating fatty acid residues   [1]

(b) Explain how the protein shown in the diagram can actively transport the glucose molecule into the cell.

   Energy is released from mitochondria near the channel protein the channel protein opens   [3]

Note the help given in diagrams. The labelling of the protein molecule may trigger the memory so that the candidate has to make a small step to link the 'channel' function to this diagram. Examiners give clues! Expect more clues at AS Level than at A2 Level.

Extended answers

In AS Biology, questions requiring more extended answers will usually form part of structured questions. They will normally appear at the end of a structured question and will typically have a value of three to six marks. Longer answers are allocated more lines, so you can use this as a guide as to the extent of your answer. The mark allocation is a guide as to how many points you need to make in your response. Often for an answer worth six marks the mark scheme could have eight creditable answers. You are awarded up to the maximum, six in this instance.

Longer extended questions may be set. These are often open response questions. These questions may be worth up to ten marks for full credit. Extended answers are used to allocate marks for the **quality of written communication**.

Candidates are assessed on their ability to use a suitable style of writing, and organise relevant material, both logically and clearly. The use of specialist biological words in context is also assessed. Spelling, punctuation and grammar are also taken into consideration. Here is a longer extended response question.

Question

Give an account of the effects of sewage entry into a river and explain the possible consequences to organisms downstream.

The sewage enters the river and is decomposed by bacteria. ✔ These bacteria are saprobiotic ✔ they produce nitrates which act as a fertiliser. ✔ Algae form a blanket on the surface ✔ light cannot reach plants under the algae so these plants die. ✔ Bacteria decompose the dead plants ✔ the bacteria use oxygen/bacteria are aerobic ✔ fish die due to lack of oxygen ✔ Tubifex worms or bloodworms increase near sewage entry ✔ mayfly larvae cannot live close to sewage entry/mayfly larvae appear a distance downstream where oxygen levels return. ✔

10 marking points → [7]

Remember that mark schemes for extended questions often exceed the question total, but you can only be awarded credit up to a maximum. Examiners sometimes build in a hurdle, e.g. in the above responses, references to one organism which increases in population is worthy of a mark, and another which decreases in population is worth another. Continually referring to different species which repeat a growth pattern will not gain further credit.

Exam technique

AS Biology builds from grade C in GCSE Science and GCSE Additional Science (combined) or GCSE Biology. This study guide has been written so that you will be able to tackle AS Biology from a GCSE science background.

You should not need to search for important Biology from GCSE science because this has been included where needed in each chapter. If you have not studied science for some time, you should still be able to learn AS Biology using this text alone.

What are examiners looking for?

Whatever type of question you are answering, it is important to respond in a suitable way. Examiners use instructions to help you to decide the length and depth of your answer. The most common words used are given below, together with a brief description of what each word is asking for.

Define

This requires a formal statement. Some definitions are easy to recall.

Define the term active transport.

This is the movement of molecules from where they are in lower concentration to where they are in higher concentration. The process requires energy.

Other definitions are more complex. Where you have problems it is helpful to give an example.

Define the term endemic.

This means that a disease is found regularly in a group of people, district or country. Use of an example clarifies the meaning. Indicating that malaria is invariably found everywhere in a country, confirms understanding.

Explain

This requires a reason. The amount of detail needed is shown by the number of marks allocated.

Explain the difference between resolution and magnification.

Resolution is the ability to be able to distinguish between two points whereas magnification is the number of times an image is bigger than an object itself.

State

This requires a brief answer without any reason.

State one role of blood plasma in a mammal.

Transport of hormones to their target organs.

List

This requires a sequence of points with no explanation.

List the abiotic factors which can affect the rate of photosynthesis in pond weed.

carbon dioxide concentration; amount of light; temperature; pH of water

Describe

This requires a piece of prose which gives key points. Diagrams should be used where possible.

Describe the nervous control of heart rate.

The medulla oblongata ✓ of the brain connects to the sino atrial node in the right atrium, wall ✓ via the vagus nerve and the sympathetic nerve ✓ the sympathetic nerve speeds up the rate ✓ the vagus nerve slows it down. ✓

Discuss

This requires points both for and against, together with a criticism of each point. (**Compare** is a similar command word.)

Discuss the advantages and disadvantages of using systemic insecticides in agriculture.

Advantages are that the insecticides kill the pests which reduce yield ✓ they enter the sap of the plants so insects which consume sap die ✓ the insecticide lasts longer than a contact insecticide, 2 weeks is not uncommon ✓

Disadvantages are that insecticide may remain in the product and harm a consumer e.g. humans ✓ it may destroy organisms other than the target ✓ no insecticide is 100% effective and develops resistant pests. ✓

Suggest

This means that there is no single correct answer. Often you are given an unfamiliar situation to analyse. The examiners hope for logical deductions from the data given and that, usually, you apply your knowledge of biological concepts and principles.

The graph shows that the population of lynx decreased in 1980. Suggest reasons for this.

Weather conditions prevented plant growth ✓ so the snowshoe hares could not get enough food and their population remained low ✓ so the lynx did not have enough hares(prey) to predate upon. ✓ The lynx could have had a disease which reduced numbers. ✓

Calculate

This requires that you work out a numerical answer. Remember to give the units and to show your working, marks are usually available for a partially correct answer. If you work everything out in stages write down the sequence. Otherwise if you merely give the answer and it is wrong, then the working marks are not available to you.

Calculate the Rf value of spot X. (X is 25 mm from start and solvent front is 100 mm)

$$Rf = \frac{\text{distance moved by spot}}{\text{distance moved by the solvent front}}$$

$$= \frac{25 \text{ mm}}{100 \text{ mm}}$$

$$= 0.25$$

Outline

This requires that you give only the main points. The marks allocated will guide you on the number of points which you need to make.

Outline the use of restriction endonuclease in genetic engineering.

The enzyme is used to cut the DNA of the donor cell. ✓

It cuts the DNA up like this A T | G C C G A T = A T   + G C C G A T ✓
                          T A C G G C | T A   T A C G G C   T A

The DNA in a bacterial plasmid is cut with the same restriction endonuclease. ✓

The donor DNA will fit onto the sticky ends of the broken plasmid. ✓

If a question does not seem to make sense, you may have mis-read it. Read it again!

Some dos and don'ts

Dos

Do answer the question

No credit can be given for good Biology that is irrelevant to the question.

Do use the mark allocation to guide how much you write

Two marks are awarded for two valid points – writing more will rarely gain more credit and could mean wasted time or even contradicting earlier valid points.

Do use diagrams, equations and tables in your responses

Even in 'essay style' questions, these offer an excellent way of communicating biology.

Do write legibly

An examiner cannot give marks if the answer cannot be read.

Do write using correct spelling and grammar. Structure longer essays carefully

Marks are now awarded for the quality of your language in exams.

Don'ts

Don't fill up any blank space on a paper

In structured questions, the number of dotted lines should guide the length of your answer.

If you write too much, you waste time and may not finish the exam paper. You also risk contradicting yourself.

Don't write out the question again

This wastes time. The marks are for the answer!

Don't contradict yourself

The examiner cannot be expected to choose which answer is intended. You could lose a hard-earned mark.

Don't spend too much time on a part that you find difficult

You may not have enough time to complete the exam. You can always return to a difficult calculation if you have time at the end of the exam.

What grade do you want?

Everyone would like to improve their grades but you will only manage this with a lot of hard work and determination. You should have a fair idea of your natural ability and likely grade in biology and the hints below offer advice on improving that grade.

For a Grade A

You will need to be a very good all-rounder.

- You must go into every exam knowing the work extremely well.
- You must be able to apply your knowledge to new, unfamiliar situations.
- You need to have practised many, many exam questions so that you are ready for the type of question that will appear.

The exams test all areas of the syllabus and any weaknesses in your biology will be found out. There must be no holes in your knowledge and understanding. For a Grade A, you must be competent in all areas.

For a Grade C

You must have a reasonable grasp of biology but you may have weaknesses in several areas and you will be unsure of some of the reasons for the biology.

- Many Grade C candidates are just as good at answering questions as the Grade A students but holes and weaknesses often show up in just some topics.
- To improve, you will need to master your weaknesses and you must prepare thoroughly for the exam. You must become a better all-rounder.

For a Grade E

You cannot afford to miss the easy marks. Even if you find biology difficult to understand and would be happy with a Grade E, there are plenty of questions in which you can gain marks.

- You must memorise all definitions.
- You must practise exam questions to give yourself confidence that you do know some biology. In exams, answer the parts of questions that you know first. You must not waste time on the difficult parts. You can always go back to these later.
- The areas of biology that you find most difficult are going to be hard to score on in exams. Even in the difficult questions, there are still marks to be gained. Show your working in calculations because credit is given for a sound method. You can always gain some marks if you get part of the way towards the solution.

What marks do you need?

The table below shows how your average mark is transferred into a grade.

average	80%	70%	60%	50%	40%
grade	A	B	C	D	E

To achieve an A* grade, you will need to achieve a...
- grade A overall (80% or more on uniform mark scale) for the **whole** A level qualification
- grade A* (90% or more on the uniform mark scale) across your A2 units.

A* grades are awarded for the A level qualification only and not for the AS qualification or individual units

Four steps to successful revision

Step 1: Understand

- Study the topic to be learned slowly. Make sure you understand the logic or important concepts.
- Mark up the text if necessary – underline, highlight and make notes.
- Re-read each paragraph slowly.

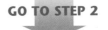

GO TO STEP 2

Step 2: Summarise

- Now make your own revision note summary:
  What is the main idea, theme or concept to be learnt?
  What are the main points? How does the logic develop?
  Ask questions: Why? How? What next?
- Use bullet points, mind maps, patterned notes.
- Link ideas with mnemonics, mind maps, crazy stories.
- Note the title and date of the revision notes
  (e.g. Biology: Cells, 3rd March).
- Organise your notes carefully and keep them in a file.

This is now in **short-term memory**. You will forget 80% of it if you do not go to Step 3.
GO TO STEP 3, but first take a 10 minute break.

Step 3: Memorise

- Take 25 minute learning 'bites' with 5 minute breaks.
- After each 5 minute break test yourself:
  Cover the original revision note summary.
  Write down the main points.
  Speak out loud (record on tape).
  Tell someone else.
  Repeat many times.

The material is well on its way to **long-term memory**.
You will forget 40% if you do not do step 4. **GO TO STEP 4**

Step 4: Track/Review

- Create a Revision Diary (one A4 page per day).
- Make a revision plan for the topic, e.g. 1 day later, 1 week later, 1 month later.
- Record your revision in your Revision Diary, e.g.
  Biology: Cells, 3rd March 25 minutes
  Biology: Cells, 5th March 15 minutes
  Biology: Cells, 3rd April 15 minutes
  ... and then at monthly intervals.

Chapter 1

Biological molecules

The following topics are covered in this chapter:

- Carbohydrates
- Lipids

- Proteins
- The importance of water to life

1.1 Carbohydrates

After studying this section you should be able to:

- recall the main elements found in carbohydrates
- recall the structure of glucose, fructose, lactose, sucrose, starch, glycogen and cellulose
- recall the role of glucose, starch, glycogen, cellulose and pectin

LEARNING SUMMARY

Structure of carbohydrates

EDEXCEL   1.1.4
           2.4.3

Monosaccharides

All carbohydrates are formed from the elements carbon (C), hydrogen (H) and oxygen (O). The formula of a carbohydrate is always $(CH_2O)_n$. The n represents the number of times the basic CH_2O unit is repeated, e.g. where n = 6 the molecular formula is $C_6H_{12}O_6$. This is the formula shared by glucose and other simple sugars like fructose. These simple sugars are made up from a single sugar unit and are known as **monosaccharides**.

The molecular formula, $C_6H_{12}O_6$, does not indicate how the atoms bond together. Bonded to the carbon atoms are a number of $-H$ and $-OH$ groups. Different positions of these groups on the carbon chain are responsible for different properties of the molecules. The structural formulae of α and β glucose are shown below.

> These molecules are mirror images of each other. When molecules have the same molecular formula but different structural formulae, they are known as **isomers**. Isomers have different properties to each other.

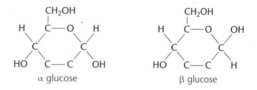

Glucose is so small that it can pass through the villi and capillaries into our bloodstream. The molecules subsequently release energy as a result of respiration. Simple glucose molecules are capable of so much more than just releasing energy. They can combine with others to form bigger molecules.

Disaccharides

Each glucose unit is known as a **monomer** and is capable of linking others. This diagram shows two molecules of α glucose forming a **disaccharide**.

> In your examinations look for different monosaccharides being given, like fructose or β glucose. You may be asked to show how they bond together. The principle will be exactly the same.

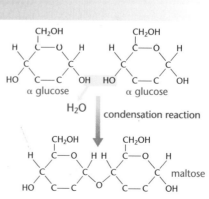

A **condensation** reaction means that as two carbohydrate molecules bond together a water molecule is produced. The link formed between the two glucose molecules is known as a **glycosidic bond**.

A glycosidic bond can also be broken down to release separate monomer units. This is the opposite of the reaction shown above. Instead of water being given off, a water molecule is needed to break each glycosidic bond. This is called **hydrolysis** because water is needed to split up the bigger molecule.

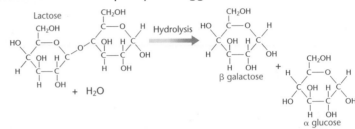

'**Lysis**' literally means '**splitting**'. In hydrolysis water is needed in the reaction to break down the molecule.

Different disaccharides are made by joining together different monosaccharides.

Disaccharide	Component monosaccharides
lactose	glucose + galactose
maltose	glucose + glucose
sucrose	glucose + fructose

Polysaccharides

Like disaccharides, **polysaccharides** consist of monomer units linked by the glycosidic bond. However, instead of just two monomer units they can have many. Chains of these 'sugar' units are known as **polymers**. These larger molecules have important structural and storage roles.

Starch is a polymer of the sugar, α glucose. The diagram below shows part of a starch molecule.

Notice the five glycosidic bonds on just a small part of a starch molecule.

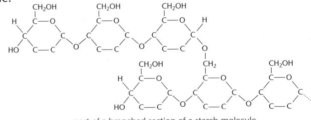

part of a branched section of a starch molecule

This type of starch molecule is called **amylopectin** and it has a branched structure. Starch also contains **amylose**. This does not contain branches but the chain of glucose units forms a helix. **Glycogen** is similar in structure to amylopectin but with more branches. Cellulose is also a polymer of glucose units, but this time the units are β glucose.

How useful are polysaccharides?

- **Starch** is stored in organisms as a future energy source, e.g. potato has a high starch content to supply energy for the buds to grow at a later stage.
- **Glycogen** is stored in the liver, which releases glucose for energy in times of low blood sugar.

Both starch and glycogen are insoluble which enables them to remain inside cells.

The many branches in the amylopectin molecule means that enzymes can digest the molecule rapidly.

- **Cellulose** has long molecules which help form a tough protective layer around plant cells, the cell wall. Each cellulose molecule has up to 10 000 β glucose units. Each molecule can form cross-links with other cellulose molecules forming fibres. This makes cellulose fibres very strong.

Pectins help cells to bind together.

- **Pectins** are used alongside cellulose in the cell wall. They are polysaccharides which are bound together by calcium pectate.

Together the cellulose and pectins give exceptional mechanical strength. The cell wall is also permeable to a wide range of substances.

1.2 Lipids

After studying this section you should be able to:

- recall the main elements found in lipids
- recall the structure of lipids
- distinguish between saturated and unsaturated fats
- recall the role of lipids

LEARNING SUMMARY

What are lipids?

EDEXCEL   1.1.5

Lipids include **oils**, **fats** and **waxes**. They consist of exactly the same elements as carbohydrates, i.e. carbon (C), hydrogen (H) and oxygen (O), but their proportion is different. There is always a high proportion of carbon and hydrogen, with a small proportion of oxygen.

The diagram below shows the structural formula of the most common type of lipid called a **triglyceride**.

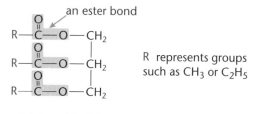

R represents groups such as CH_3 or C_2H_5

a **triglyceride** fat

Triglycerides are formed when fatty acids react with glycerol. During this reaction water is produced, a further example of a condensation reaction. The essential bond is the **ester bond**.

Note that water is produced during triglyceride formation. This is another example of a condensation reaction. Different triglyceride fats are formed from different fatty acids.

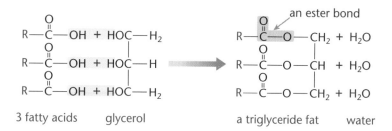

3 fatty acids   glycerol   a triglyceride fat   water

Triglycerides can be changed back into the original fatty acids and glycerol. Enzymes are needed for this transformation together with water molecules. Remember, an enzyme reaction which requires water to break up a molecule is known as **hydrolysis**.

What are saturated and unsaturated fats?

The answer lies in the types of fatty acid used to produce them.

The hydrocarbon chains are so long that they are often represented by the acid group (–COOH) and a zig-zag line.

unsaturated

∿∿∿=∿∿COOH

saturated

∿∿∿∿∿∿COOH

stearic acid

a **saturated** fatty acid

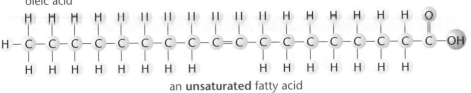

oleic acid

an **unsaturated** fatty acid

> **KEY POINT**
>
> Saturated fatty acids have no C=C (double bonds) in their hydrocarbon chain, but unsaturated fatty acids do. This is the difference.

How useful are lipids?

Like carbohydrates, lipids are used as an **energy** supply, but a given amount of lipid releases more energy than the same amount of carbohydrate. Due to their **insolubility** in water and **compact** structure, lipids have long-term **storage** qualities. Adipose cells beneath our skin contain large quantities of fat which **insulate us** and help to maintain body temperature. Fat gives **mechanical** support around our soft organs and even gives **electrical insulation** around our nerve axons.

An aquatic organism such as a dolphin has a large fat layer which:

- is an energy store
- is a thermal insulator
- helps the animal remain buoyant.

The most important role of lipids is their function in cell membranes. To fulfil these functions a triglyceride fat is first converted into a **phospholipid**.

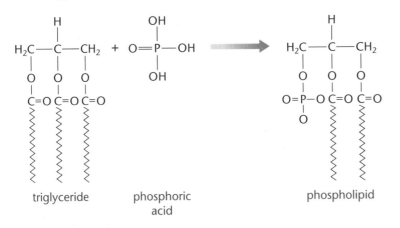

triglyceride   phosphoric acid   phospholipid

Phosphoric acid replaces one of the fatty acids of the triglyceride. The new molecule, the phospholipid, is a major component of cell membranes. Cell membranes also contain the lipid cholesterol. The diagram below represents a phospholipid.

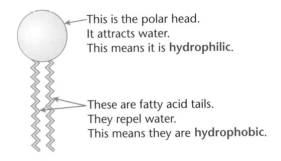

This is the polar head.
It attracts water.
This means it is **hydrophilic**.

These are fatty acid tails.
They repel water.
This means they are **hydrophobic**.

a phospholipid

1.3 Proteins

The building blocks of proteins

EDEXCEL 1.2.7

Like carbohydrates and lipids, proteins contain the elements carbon (C), hydrogen (H) and oxygen (O), but in addition they **always** contain **nitrogen** (N). Sulfur is also often present.

Before understanding how proteins are constructed, the structure of **amino acids** should be noted. The diagram below shows the general structure of an amino acid.

> Just like the earlier carbohydrate and lipid molecules, 'R' represents groups such as $-CH_3$ and $-C_2H_5$. There are about 20 commonly found amino acids but you will not need to know them all. Instead, learn the basic structure shown opposite.

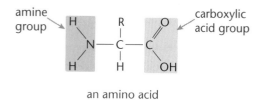

an amino acid

How is a protein constructed?

The process begins by amino acids bonding together. The diagram shows two amino acids being joined together by a **peptide bond**.

> This is another example of a condensation reaction as water is produced as the dipeptide molecule is assembled.
>
> Note that the peptide bonds can be broken down by a hydrolysis reaction.

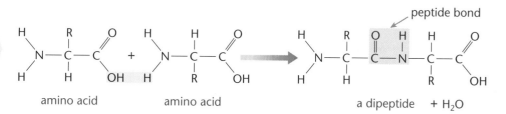

amino acid   amino acid   a dipeptide $+ H_2O$

> The sequence of amino acids along a polypeptide is controlled by another complex molecule, DNA (see the genetic code, page 54).

When many amino acids join together a long-chain **polypeptide** is produced. The linking of amino acids in this way takes place during protein synthesis (see page 55). There are around 20 different amino acids. Organisms join amino acids in different linear sequences to form a variety of polypeptides, then build these polypeptides into complex molecules, the **proteins**. Humans need eight essential amino acids as adults and ten as children, all the others can be made inside the cells.

Levels of organisation in proteins

EDEXCEL 1.2.7

Primary protein structure

This is the **linear sequence** of amino acids.

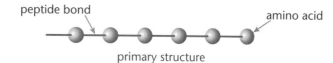

primary structure

Secondary protein structure

Polypeptides become twisted or coiled. These shapes are known as the **secondary structure**. There are two common secondary structures; the α-helix and the β-pleated sheet.

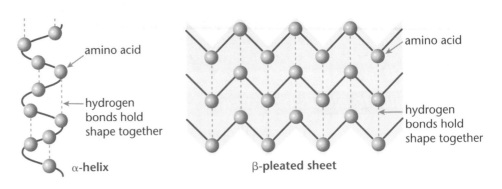

The polypeptides are held in position by hydrogen bonds. In both α-helices and β-pleated sheets the **C = O** of one amino acid bonds to the **H–N** of an adjacent amino acid like this: C = O --- H–N.

$$----- = \text{hydrogen bonds}$$

An α-helix is a tight, twisted strand; a β-pleated sheet is where a zig-zag line of amino acids bonds with the next, and so on. This forms a sheet or ribbon shape.

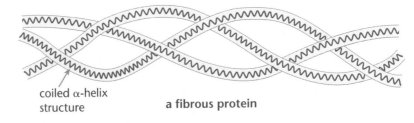

coiled α-helix structure

a fibrous protein

Tertiary protein structure

This is when a polypeptide is **folded** into a **precise** shape. The polypeptide is held in 'bends' and 'tucks' in a **permanent** shape by a range of bonds including:

- **disulfide** bridges (sulfur–sulfur bonds)
- hydrogen bonds
- ionic bonds
- hydrophobic and hydrophilic interactions.

Some proteins are folded up into a spherical shape. They are called **globular proteins**. They are soluble in water. Other proteins, called **fibrous proteins**, form long chains. They are insoluble and usually perform structural functions.

Quaternary protein structure

Some proteins consist of **different polypeptides** bonded together to form extremely intricate shapes. A haemoglobin molecule is formed from four separate polypeptide chains. It also has a haem group, which contains iron. This inorganic group is known as a **prosthetic group** and in this instance aids oxygen transport.

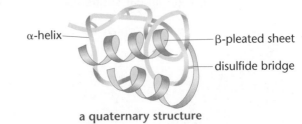

α-helix — β-pleated sheet — disulfide bridge

a quaternary structure

Both secondary structures give additional strength to proteins. The α-helix helps make tough fibres like the protein in your nails, e.g. keratin. The β-pleated sheet helps make the strength-giving protein in silk, fibroin. Many proteins are made from both α-helix and β-pleated sheet.

The protein shown only achieves a secondary structure as the simple α-helix polypeptides do not undergo further folding.

This is the structure of a fibrous protein. It is made of three α-helix polypeptides twisted together.

Note that the specific contours of proteins have extremely significant roles in life processes. (See enzymes page 34.)

This is the structure of a **globular** protein. It is made of an α-helix and a β-pleated sheet. Precise shapes are formed with specific contours.

Note that some proteins do not have a quaternary structure. If they consist of just one folded polypeptide then they are classified as having tertiary structure. If they are simple fibres of α-helices or β-pleated sheets then they have only secondary protein structure.

How useful are proteins?

EDEXCEL 1.2.7

Proteins can be just as beneficial as carbohydrates and lipids in releasing energy. Broken down into their component amino acids, they liberate energy during respiration. The list below shows **important** uses of proteins:

- **cell-membrane proteins** transport substances across the membrane for processes such as facilitated diffusion and active transport

- **enzymes** catalyse biochemical reactions, e.g. pepsin breaks down protein into polypeptides

- **hormones** are passed through the blood and trigger reactions in other parts of the body, e.g. insulin regulates blood sugar

- **immuno-proteins**, e.g. antibodies are made by lymphocytes and act against antigenic sites on micro-organisms

- **structural proteins** give strength to organs, e.g. collagen makes tendons tough

- **transport proteins**, e.g. haemoglobin transports oxygen in the blood

- **contractile proteins**, e.g. actin and myosin help muscles shorten during contraction

- **storage proteins**, e.g. aleurone in seeds helps germination, and casein in milk helps supply valuable protein to babies

- **buffer proteins**, e.g. blood proteins, due to their charge, help maintain the pH of plasma.

Progress check

1 List the sequence of structures in a globular protein such as haemoglobin.

2 The following statements refer to proteins used for different functions in the body. The list gives the name of different types of protein. Match the name of each type of protein with the correct statement.

(i) transport proteins   (vi) contractile proteins
(ii) immuno-proteins   (vii) enzymes
(iii) storage proteins   (viii) structural proteins
(iv) buffer proteins   (ix) hormones
(v) cell-membrane proteins

A haemoglobin is used to transport oxygen in blood.

B aleurone in seeds is a source of amino acids as it is broken down during germination.

C actin and myosin help muscles shorten during contraction.

D antibodies made by lymphocytes against antigens.

E blood proteins, due to their charge, help maintain the pH of plasma.

F used to transport substances across the membrane for processes such as facilitated diffusion.

G passed through blood, used to trigger reactions in other parts of the body, e.g. FSH stimulates a primary follicle.

H used to catalyse biochemical reactions, e.g. amylase breaks down starch into maltose.

I used to give strength to organs, e.g. collagen makes tendons tough.

2 A (i), B (iii), C (vi), D (ii), E (iv), F (v), G (ix), H (vii), I (viii).

1 primary structure: amino acids linked in a linear sequence; secondary structure: α-helix or β-pleated sheet; tertiary structure: further folding of polypeptide held by disulfide bridges, ionic bonds, and hydrogen bonds; quaternary structure: two or more polypeptides bonded together.

1.4 The importance of water to life

After studying this section you should be able to:

- recall the properties of water
- recall the functions of water

Properties and uses of water

EDEXCEL   1.1.2

Water is essential to living organisms. The list below shows some of its properties and uses.

- **Hydrogen bonds** are formed between the oxygen of one water molecule and the hydrogen of another. As a result of this water molecules have an attraction for each other known as **cohesion**.

- **Cohesion** is responsible for surface tension which enables aquatic insects like pond skaters to walk on a pond surface. It also aids capillarity, the way in which water moves through xylem in plants.

- Water is a **dipolar** molecule, which means that the oxygen has a slight negative charge at one end of the molecule, and each hydrogen has a slight positive charge at the other end.

- Other **polar** molecules dissolve in water. The different charges on these molecules enable them to fit into water's hydrogen bond structure. Ions in solution can be transported or can take part in reactions. Polar substances can dissolve in water and are called **hydrophilic**. Non-polar substances cannot dissolve in water and are **hydrophobic**.

- Water is used in **photosynthesis**, so it is necessary for the production of glucose. This in turn is used in the synthesis of many chemicals.

- Water helps in the **temperature regulation** of many organisms. It enables the cooling down of some organisms. Owing to a **high latent heat of vaporisation**, large amounts of body heat are needed to evaporate a small quantity of water. Organisms like humans cool down effectively but lose only a small amount of water in doing so.

- A relatively high level of heat is needed to raise the temperature of water by a small amount due to its **high specific heat capacity**. This enables organisms to control their body temperature more effectively.

- Water is a solvent for ionic compounds. A number of the essential elements required by organisms are obtained in ionic form, e.g.:
  (a) plants absorb nitrate ions (NO_3^-) and phosphate ions (PO_4^-) in solution
  (b) animals intake sodium ions (Na^+) and chloride ions (Cl^-).

Try to learn all of the functions of water molecules given in the list. Water is used in so many ways that the chance of being questioned on the topic is high.

Sample questions and model answers

1 Below are the structures of two glucose molecules.

(a) Complete the equation to show how the molecules react to form a glycosidic bond and the molecule produced.

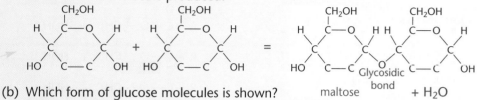

(b) Which form of glucose molecules is shown? Give a reason for your answer. [2]

α glucose, because the −OH groups on carbon atom 1 are down

(c) State the type of reaction which takes place when the two molecules shown above react together. [2]

Condensation.

2 The diagram below shows a globular protein consisting of four polypeptide chains.

α-helix

β-pleated sheet

disulfide bridge

(a) Use your own knowledge and the information given to explain how this protein shows primary, secondary, tertiary and quaternary structure. [5]

Primary structure: it is formed from chains of amino acids; it has polypeptides made of a linear sequence of amino acids.
Secondary structure: it has an α-helix, it has a β-pleated sheet.
Tertiary structure: the polypeptides are folded, the folds are held in position by disulfide bridges.
Quaternary structure: there are four polypeptides in this protein.
Two or more are bonded together to give a quaternary structure.

(b) Name and describe a test which would show that haemoglobin is a protein. [3]
The Biuret test.
Take a sample of haemoglobin and add water, sodium hydroxide and copper sulfate.
The colour of the mixture shows as violet or mauve if the sample is a protein.

Practice examination questions

Try all of the questions and check your answers with the mark scheme on page 74.

1 (a) Complete the equation below to show the breakdown of a triglyceride fat into fatty acids and glycerol. [2]

a triglyceride fat   3 molecules of water

(b) Describe a biochemical test which would show if a sample contained fat. [3]

2 The diagram below shows a polypeptide consisting of 15 amino acids.

(a) Name the bond between each pair of amino acids in this polypeptide. [1]

(b) What is group X? [1]

(c) Which level of protein structure is shown by this polypeptide? Give a reason for your answer. [2]

3 Explain how the following properties of water are useful to living organisms:

(a) a large latent heat of evaporation [2]

(b) a high specific heat capacity [2]

(c) the cohesive attraction of water molecules for each other. [2]

Chapter 2
Cells

The following topics are covered in this chapter:

- *The ultra-structure of cells*
- *Specialisation of cells*

2.1 The ultra-structure of cells

After studying this section you should be able to:

- *identify cell organelles and understand their roles*
- *recall the differences between prokaryotic and eukaryotic cells*

Cell organelles

EDEXCEL 2.3.3 – 4
2.4.4

The cell is the basic functioning unit of organisms in which chemical reactions take place. These reactions involve energy release needed to support life and build structures. Organisms consist of one or more cells. The amoeba is composed of one cell, whereas millions of cells make up a human.

> **KEY POINT**
> Every cell possesses internal coded instructions to control cell activities and development. Cells also have the ability to continue life by some form of cell division.

Organelles are best seen with the aid of an electron microscope.

The ultra-structure of a cell can be seen using an electron microscope. Sub-cellular units called **organelles** become visible. Each organelle has been researched to help us understand more about the processes of life.

The animal cell and its organelles

The diagram below shows the organelles found in a typical animal cell.

A plant cell has all of the same structures plus:
- *a cellulose cell wall*
- *chloroplasts (some cells)*
- *a sap vacuole with tonoplast.*

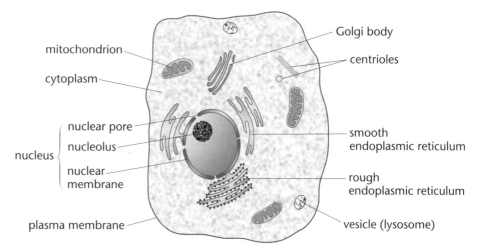

mitochondrion — cytoplasm — nuclear pore — nucleolus — nuclear membrane — nucleus — plasma membrane — Golgi body — centrioles — smooth endoplasmic reticulum — rough endoplasmic reticulum — vesicle (lysosome)

Cell surface (plasma) membrane

This covers the outside of a cell and consists of a **double** layered sheet of phospholipid molecules interspersed with proteins. It separates the cell from the outside environment, gives physical protection and allows the import and export of **selected** chemicals.

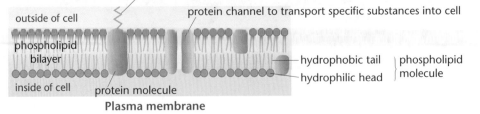

carbohydrate chain

outside of cell

protein channel to transport specific substances into cell

phospholipid bilayer

hydrophobic tail ⎱ phospholipid
hydrophilic head ⎰ molecule

inside of cell

protein molecule

Plasma membrane

Nucleus

This controls all cellular activity using coded instructions located in DNA. These coded instructions enable the cell to make specific proteins. RNA is produced in the nucleus and leaves via the nuclear pores. The nucleus stores, replicates and decodes DNA.

> Be ready to identify all cell organelles in either a diagram or electron micrograph. A mitochondrion is often sausage shaped but the end view is circular. Look out for the internal membranes.

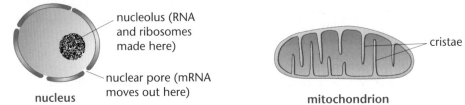

nucleolus (RNA and ribosomes made here)

nuclear pore (mRNA moves out here)

nucleus

cristae

mitochondrion

Mitochondria

These consist of a double membrane enclosing a semi-fluid matrix. Throughout the matrix is an internal membrane, folded into cristae. The cristae and matrix contain enzymes which enable this organelle to carry out aerobic respiration. It is the key organelle in the release of energy, making ATP available to the cell.

Mitochondria are needed for many energy requiring processes in the cell, including active transport and the movement of cilia.

Cytoplasm

> Cytoplasm is often seen as grey and granular. If the image is 'clear' then you are probably looking at a vacuole.

Each organelle in a cell is suspended in a semi-liquid medium, the cytoplasm. Many ions are dissolved in it. It is the site of many chemical reactions.

Ribosomes

> Look for tiny dots in the cytoplasm. They will almost certainly be ribosomes. A membrane adjacent to a line of ribosomes is probably the rough endoplasmic reticulum.

There are numerous **ribosomes** in a cell, located along **rough endoplasmic reticulum**. They aid the manufacture of proteins, being the site where mRNA meets tRNA so that amino acids are bonded together.

Endoplasmic reticulum (ER)

This is found as **rough ER** (with ribosomes) and **smooth ER** (without ribosomes). It is a series of folded internal membranes. Substances are transported in the spaces between the ER. The smooth ER aids the synthesis and transport of lipids.

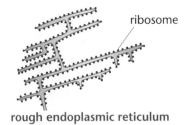

ribosome

rough endoplasmic reticulum

smooth endoplasmic reticulum

Golgi body

> Look for vesicles 'pinching off' the main Golgi sacs.

This is a series of flattened sacs, each separated from the cytoplasm by a membrane. The **Golgi body** is a packaging system where important chemicals become membrane wrapped, forming **vesicles**. The vesicles become detached from the main Golgi sacs, enabling the isolation of chemicals from each other in the cytoplasm. The Golgi body aids the production and secretion of many proteins, carbohydrates and glycoproteins. Vesicle membranes merge with the plasma membrane to enable secretions to take place.

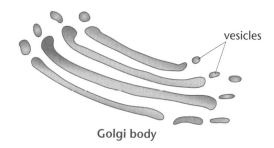

Golgi body

Lysosomes

These are **specialised vesicles** because they contain **digestive enzymes**. The enzymes have the ability to break down proteins and lipids. If the enzymes were free to react in the cytoplasm then cell destruction would result.

lysosome

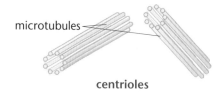

microtubules

centrioles

Centrioles

In a cell there are two short cylinders which contain **microtubules**. Their function is to aid cell division. During division they move to opposite poles as the spindle develops.

The plant cell and its organelles

All of the structures described for animal cells are also found in plant cells (except the centrioles). Additionally there are three extra structures shown in the diagram below.

Did you spot the three extra structures in the plant cell? Remember that a root cell under the soil will not possess chloroplasts. Nor does every plant cell above the soil have chloroplasts.

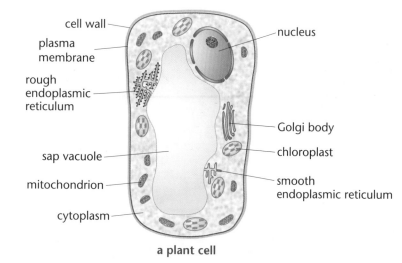

cell wall

plasma membrane

rough endoplasmic reticulum

sap vacuole

mitochondrion

cytoplasm

nucleus

Golgi body

chloroplast

smooth endoplasmic reticulum

a plant cell

Cell wall

Cells other than plant cells can have cell walls, e.g. bacteria have polysaccharides other than cellulose.

Around the plasma membrane of plant cells is the cell wall. This is secreted by the cell and consists of cellulose microfibrils embedded in a layer of calcium pectate and hemicelluloses. Between the walls of neighbouring cells calcium pectate cements one cell to the next in multi-cellular plants. Plant cell walls provide a rigid support for the cell but allow many substances to be imported or exported by the cell. The wall allows the cell to build up an effective hydrostatic skeleton. Some plant cells have a cytoplasmic link which crosses the wall. These links of cytoplasm are known as **plasmodesmata**.

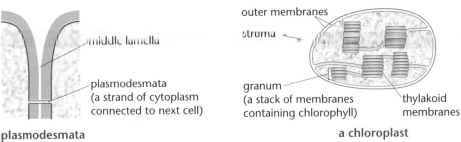

plasmodesmata

a chloroplast

Chloroplasts

These enable the plant to photosynthesise, making glucose. Each consists of an outer covering of two membranes. Inside are more membranes stacked in piles called **grana**. The membranes enclose a substance vital to photosynthesis, **chlorophyll**. Inside the chloroplast is a matrix known as the **stroma** which is also involved in photosynthesis.

Sap vacuole

This is a large space in a plant cell, containing chemicals such as glucose and mineral ions in water. This solution is the sap. It is surrounded by a membrane known as the **tonoplast**. It is important that a plant cell contains enough water to maintain internal hydrostatic pressure. When this is achieved the cell is turgid, having maximum hydrostatic strength.

Prokaryotic and eukaryotic cells

EDEXCEL   2.3.2

Organisms can be classified into two groups, **prokaryotic** or **eukaryotic** according to their cellular structure.

> **KEY POINT**
>
> Prokaryotic cells are characteristic of two groups of organisms, bacteria and blue-green algae. Prokaryotic cells are less complex than eukaryotic ones and are considered to have evolved earlier.

The table below states similarities and differences between the two types of organism.

In an examination you will often be given a diagram of a cell from an organism you have not seen before. This is not a problem! The examiners are testing your recognition of the organelles found in typical prokaryotic and eukaryotic organisms.

		Prokaryotic cells	Eukaryotic cells
Kingdom		Prokaryotae	Protoctista, Fungi, Animalia, Plantae
Organelles	1	small ribosomes	large ribosomes
	2	DNA present but there is no nuclear membrane	DNA is enclosed in a membrane i.e. has nucleus, mitochondria, (Golgi body vesicles and ER are present)
	3	cell wall present consisting of mucopeptides	cell walls present in plant cells – cellulose cell walls present in fungi – chitin
	4	if cells have flagellae there is no 9+2 microtubule arrangement	if cells have flagellae there is a 9+2 microtubule arrangement

Progress check

1 Describe the function of each of the following cell organelles:

   nucleus   centrioles   Golgi body
   mitochondria   ribosomes   cell (plasma) membrane

2 Give **three** structural differences between a plant and animal cell.

2 A plant cell has a cellulose cell wall, chloroplasts, and a sap vacuole lined by a tonoplast.

Cell (plasma) membrane – gives physical protection to the outside of a cell, allows the import and export of *selected chemicals*.
Golgi body – is a packaging system where chemicals become membrane wrapped, forming vesicles
ribosomes – aid the manufacture of proteins, being the site where mRNA meets tRNA so that amino acids are bonded together
centrioles – help produce the spindle during cell division
mitochondria – release energy during aerobic respiration
1 nucleus – mRNA is produced in the nucleus with the help of DNA

2.2 Specialisation of cells

After studying this section you should be able to:

● *understand how cells aggregate into tissues and organs*

The earlier parts of this chapter described the structure and function of generalised animal and plant cells. Their features are found in many unicellular organisms where all the life-giving processes are carried out in one cell. Additionally many multicellular organisms exist. A few show no specialisation and consist of repeated identical cells, e.g. Volvox, a colonial alga. Most multicellular organisms exhibit **specialisation**, where different cells are adapted for specific roles.

Tissues, organs and systems

EDEXCEL   2.3.5

A **tissue** is a collection of similar cells, derived from the same source, all working together for a specific function, e.g. palisade cells of the leaf which photosynthesise or the smooth muscle cells of the intestine which carry out peristalsis.

Two examples of plant tissues are **xylem** and **sclerenchyma**. One part of the plant where they are both found is in the stem. They are found in collections of tissues called vascular bundles.

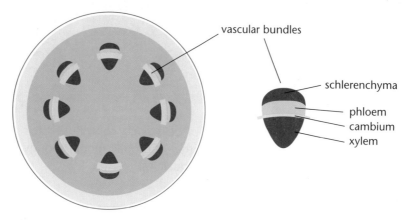

vascular bundles

schlerenchyma

phloem
cambium
xylem

These two tissues have different structures and functions.

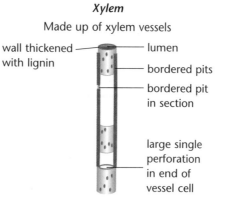

Xylem
Made up of xylem vessels

wall thickened with lignin

lumen

bordered pits

bordered pit in section

large single perforation in end of vessel cell

used to support the stem and to transport water and mineral ions up to the leaves

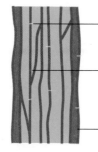

Sclerenchyma
made up off sclerenchyma fibres

pits connect the fibres together

pointed ends of cells locked together

wall strengthened with lignin

supports the stem

Water is needed by the leaves for photosynthesis and to support the cells. Mineral ions are needed for various functions, e.g. nitrates to make amino acids, magnesium for chlorophyll and calcium to make calcium pectate.

An **organ** is a collection of tissues which combine their properties for a specific function, e.g. the stomach includes: smooth muscle, epithelial lining cells, connective tissue, etc. Together they enable the stomach to digest food.

A range of tissues and organs combine to form a **system**, e.g. the respiratory system.

In multicellular organisms specific groups of cells are specialised for a particular role. This increased efficiency helps the organism to have better survival qualities in the environment.

The photomicrograph below shows some of the cells which are part of a bone.

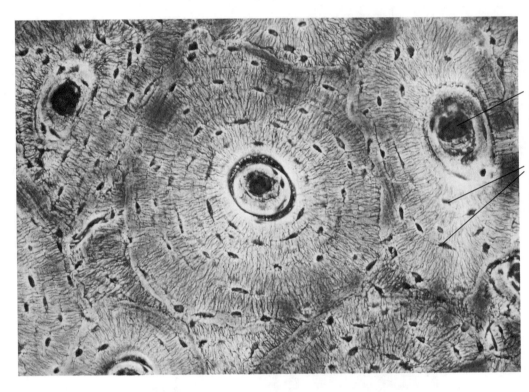

Haversian canal containing blood vessels and nerves

Bone cells which secrete the minerals which harden the bone

Combinations of cells each contribute their specific adaptations to the overall function of an organ. Compact bone, spongy bone and articular cartilage all have distinct but vital qualities.

Practice examination questions

1 The diagram shows the structure of a cell surface membrane.

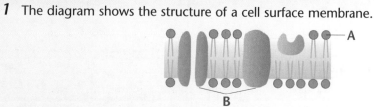

(a) Name molecules A and B. [2]

(b) Describe one role of molecule B. [1]

(c) Explain why the structure of molecule A means that the membrane forms a bilayer [2]

2 The diagram below shows an electron micrograph of a cell.

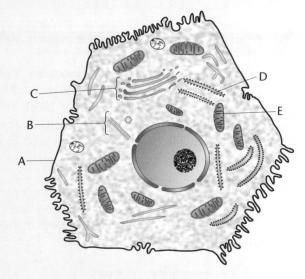

(a) Name the parts labelled in **A, B, C, D** and **E**. [5]

(b) The magnification of this diagram is 10 000.
Work out the actual diameter of the nucleus.
Give your answer in micrometers. [3]

(c) The cell is a liver cell. It contains many mitochondria.

Explain why there are so many mitochondria in each liver cell. [2]

3 The diagram below shows a bacterium.

(a) Describe **two** features, visible in the diagram, which show that the bacterium is a prokaryotic organism. [2]

(b) Name **two** organelles from a human cell which show that it is a eukaryotic organism. [2]

Chapter 3
Enzymes

The following topics are covered in this chapter:

- *Enzymes in action*

3.1 Enzymes in action

After studying this section you should be able to:

- *understand the role of the active site and the enzyme–substrate complex in enzyme action*
- *understand how enzymes catalyse biochemical reactions by lowering activation energy*
- *understand the factors which affect the rate of enzyme catalysed reactions*

LEARNING SUMMARY

How enzymes work

EDEXCEL 1.2.8-9

Living cells carry out many biochemical reactions. These reactions take place **rapidly** due to the presence of enzymes. All enzymes consist of **globular proteins** which have the ability to 'drive' biochemical reactions. Some enzymes require additional non-protein groups to enable them to work efficiently, e.g. the enzyme dehydrogenase needs a coenzyme NAD to function. Most enzymes are contained within cells but some may be released and act extracellularly.

> The tertiary folding of polypeptides are responsible for the special shape of the active site.

KEY POINT

The ability of an enzyme to function depends on the specific shape of the protein molecule. The intricate shape created by polypeptide folding (see page 22) is a key factor in both theories of enzyme action.

> In an examination the lock and key theory is the most important model to consider. Remember that both catabolic and anabolic reactions may be given.

Lock and key theory

- Some part of the enzyme has a cavity with a precise shape (**active site**).
- A substrate can fit into the active site.
- The active site (lock) is exactly the correct shape to fit the substrate (key).
- The substrate binds to the enzyme forming an **enzyme–substrate complex**.
- The reaction takes place rapidly.
- Certain enzymes break a substrate down into two or more products (**catabolic** reaction).
- Other enzymes bind two or more substrates together to assemble one product (**anabolic** reaction).

> metabolic reaction
> = anabolic + catabolic
>   reaction   reaction
>
> Remember that metabolism is a summary of **build up** and **break down** reactions.

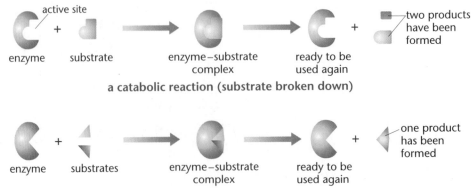

a catabolic reaction (substrate broken down)

an anabolic reaction (substrates used to build a new molecule)

Induced fit theory

- The active site is a cavity of a particular shape.
- Initially the active site is not the correct shape in which to fit the substrate.
- As the substrate approaches the active site, the site changes and this results in it being a perfect fit.
- After the reaction has taken place, and the products have gone, the active site returns to its normal shape.

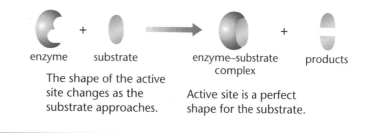

The shape of the active site changes as the substrate approaches.

Active site is a perfect shape for the substrate.

Lowering of activation energy

Every reaction requires the input of energy. Enzymes reduce the level of activation energy needed as shown by the graph.

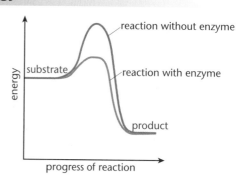

The higher the activation energy the slower the reaction. An enzyme reduces the amount of energy required for a biochemical reaction. When an enzyme binds with a substrate the available energy has a greater effect and the rate of catalysis increases. The conditions which exist during a reaction are very important when considering the rate of progress. Each of the following has an effect on the rate:

- concentration of substrate molecules
- concentration of enzyme molecules
- temperature
- pH.

You may be questioned on the factors which affect the rate of reaction. Less able candidates tend to remember just one or two factors. Learn all four factors here and achieve a higher grade!

What is the effect of enzyme and substrate concentration?

When considering the rate of an enzyme catalysed reaction the proportion of enzyme to substrate molecules should be considered. Every substrate molecule fits into an active site, then the reaction takes place. If there are more substrate molecules than enzyme molecules then the number of active sites available is a **limiting factor**. The **optimum rate** of reaction is achieved when all the active sites are in use. At this stage if more substrate is added, there is no increase in rate of product formation. When there are fewer substrate molecules than enzyme molecules the reaction will take place very quickly, as long as the conditions are appropriate.

Look out for questions which show the rate of reaction graphically. The examiners often test your understanding of limiting factors (see practice question on page 37).

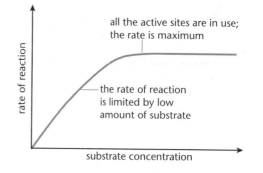

Remember that particles in liquids (and gases) are in constant random motion, even though we cannot see them.

How does temperature affect the rate of an enzyme catalysed reaction?

- **Heat energy** reaching the enzyme and substrate molecules causes them to **increase random movement**.
- The greater the heat energy the more the molecules move and so the more often they **collide**.
- The more **collisions** there are the greater the chance that substrates will fit into an **active site**, up to a specific temperature.
- At the **optimum** temperature of an enzyme, the reaction rate is maximum.
- Heat energy also affects the shape of the active site, the active site being the correct shape at the optimum temperature.
- At temperatures above optimum, the rate of reaction decreases because the active site begins to distort.
- Very high temperature causes the enzyme to become **denatured**, i.e. bonding becomes irreversibly changed and the active site is **permanently damaged**.
- At very high temperatures, the number of collisions is correspondingly high, but without active sites no products can be formed.
- At lower temperatures than the optimum, the rate of the reaction decreases because of reduced enzyme–substrate collisions.

It is interesting to consider that some micro organisms can spoil ice-cream in a freezer whereas a different micro organism, with different enzymes, can decompose grass in a 'steaming' compost heap.

Most enzymes have an optimum temperature of between 30°C and 40°C, but there are many exceptions. An example of this is shown by some bacteria that live at high temperatures in hot volcanic springs.

How does pH affect the rate of an enzyme catalysed reaction?

The **pH** of the medium can have a direct effect on the bonding responsible for the **secondary and tertiary structure** of enzymes. If the active site is changed then enzyme action will be affected. Each enzyme has an optimum pH.

Remember that other factors affect an enzyme catalysed reaction:

- substrate concentration
- enzyme concentration
- temperature.

Each can be a limiting factor.

- Many enzymes work best at **neutral** or **slightly alkaline** conditions, e.g. salivary amylase.
- Pepsin works best in **acid** conditions around pH 3.0, as expected considering that the stomach contains hydrochloric acid.

For the two enzyme examples above, the active sites are ideally shaped at the pHs mentioned. An inappropriate pH, often acidic, can change the active site drastically, so that the substrate cannot bind. The reaction will not take place. On most occasions the change of shape is not permanent and can be returned to optimum by the addition of an alkali.

Progress check

How does temperature affect the rate of the reaction by which protein is changed to polypeptides by the enzyme pepsin, in the human stomach?

$$\text{protein} \xrightarrow{\text{pepsin}} \text{polypeptides}$$

- Heat energy causes the enzyme and substrate molecules to increase random movement, increasing the chance of collision.
- At 37°C (optimum temperature) there is a greater chance that the protein will fit into an active site, so the production of polypeptides is at maximum rate.
- At 37°C (optimum temperature) the shape of the active site is best suited to fit the protein.
- At temperatures higher than 37°C the rate of reaction decreases because the active site begins to distort.
- Very high temperature causes the pepsin to become denatured, i.e. bonding has been irreversibly changed and the active site is permanently damaged.
- At very high temperatures the number of collisions is correspondingly high, but without active sites no polypeptides can be formed.
- At lower temperatures than optimum the rate of reaction decreases because of reduced enzyme–substrate collisions.

Practice examination questions

1 Describe each of the following pairs to show that you understand the main differences between them:

(a) the lock and key enzyme theory **and** the induced fit enzyme theory

(b) reversible **and** irreversible inhibitors.   [4]

2 The diagram represents an enzyme and its substrate.

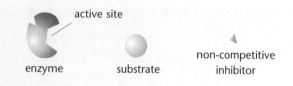

Referring to information in the diagram explain the activity of this enzyme in terms of the **induced fit theory**.   [2]

Chapter 4
Exchange

The following topics are covered in this chapter:

- The cell surface membrane
- The movement of molecules in and out of cells
- Gaseous exchange in humans
- Gaseous exchange in other organisms

4.1 The cell surface membrane

After studying this section you should be able to:

- recall the fluid mosaic model of the cell surface (plasma) membrane

LEARNING SUMMARY

Fluid mosaic model of the cell surface (plasma) membrane

EDEXCEL   1.2.2

Ultimately the exchange of substances takes place across the cell surface membrane. This must be selective, allowing some substances in and excluding others. The cell membrane consists of a bilayer of phospholipid molecules (see page 20). Each phospholipid is arranged so that the hydrophilic (attracts water) head is facing towards either the cytoplasm or the outside of the cell. The hydrophobic (repels water) tails meet in the middle of the membrane. Across this expanse of phospholipids are a number of protein molecules. Some of the proteins (intrinsic) span the complete width of the membrane, some proteins (extrinsic) are partially embedded in the membrane.

Remember that plant cells have a cellulose cell wall. This gives physical support to the cell but is permeable to many molecules. Water and ions can readily pass through.

upper surface of cell membrane protein

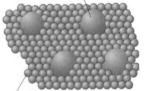

phospholipid head

The fluid mosaic model of the cell membrane

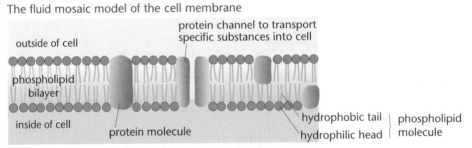

outside of cell

protein channel to transport specific substances into cell

phospholipid bilayer

inside of cell

protein molecule

hydrophobic tail
hydrophilic head

phospholipid molecule

Functions of cell membrane molecules

The term 'fluid mosaic' was given to the cell membrane because of the dynamic nature of the component molecules of the membrane. Many of the proteins seem to 'float' through an apparent 'sea' of phospholipids. Few molecules are static. The fluidity of the membrane is controlled by the quantity of cholesterol molecules. These are found inbetween the tails of the phospholipids.

Phospholipid

Small lipid-soluble molecules pass through the membrane easily because they dissolve as they pass through the phospholipid's bilayer. Small uncharged molecules also pass through the bilayer.

small lipid-soluble molecules pass through

Channel proteins (ion gates)

Larger molecules and charged molecules can pass through the membrane due to channel proteins. Some are adjacent to a receptor protein, e.g. at a synapse a transmitter substance binds to a receptor protein. This opens the channel protein or ion gate and sodium ions flow in.

Not all channel proteins need a receptor protein.

> When the molecule binds to a receptor molecule it is similar to a substrate binding with an enzyme's active site. On this occasion the receptor site is the correct shape.

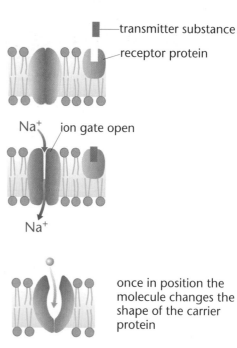

transmitter substance

receptor protein

Na^+ ion gate open

Na^+

Carrier protein molecule

Some molecules which approach a cell may bind with a carrier protein. This has a site which the incoming molecule can bind to. This causes a change of shape in the carrier protein which deposits the molecule into the cell cytoplasm.

once in position the molecule changes the shape of the carrier protein

the site gives up the molecule on the inside of the cell

carrier protein

Recognition proteins

These are extrinsic proteins, some having carbohydrate components, which help in cell recognition (cell signalling) and cell interaction, e.g. foreign protein on a bacterium would be recognised by white blood cells and the cell would be attacked. The combination of a protein with a carbohydrate is called a **glycoprotein**. The carbohydrate chains are only on the outside of the cell membrane and are called the **glycocalyx**.

> White blood cells continually check the proteins on cell membranes. Those recognised as 'self' are not attacked, whereas those which are not 'self' are attacked.

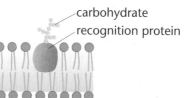

carbohydrate

recognition protein

> **KEY POINT**
>
> The cell surface membrane is the key structure which forms a barrier between the cell and its environment. Nutrients, water and ions must enter and waste molecules must leave. Equally important is the exclusion of dangerous chemicals and inclusion of vital cell contents. It is no surprise that the cell makes further compartments within the cell using membranes of similar structure to the cell surface membrane. High temperatures can destroy the structure of the cell membrane and then it loses its ability to contain the cell contents.

4.2 The movement of molecules in and out of cells

After studying this section you should be able to:

- *understand the range of methods by which molecules cross cell membranes*
- *understand the processes of diffusion, facilitated diffusion, osmosis and active transport*

LEARNING SUMMARY

How do substances cross the cell surface membrane?

EDEXCEL 1.2.3 6

Cells need to obtain substances vital in sustaining life. Some cells secrete useful substances but all cells excrete waste substances. There are several mechanisms by which molecules move across the cell surface membrane.

Diffusion

> Note that diffusion is the movement of molecules down a concentration gradient.

Molecules in liquids and gases are in constant random motion. When different concentrations are in contact, the molecules move so that they are in equal concentration throughout. An example of this is when sugar is put into a cup of tea. If left, sugar molecules will distribute themselves evenly, even without stirring. Diffusion is the movement of molecules from where they are in high concentration to where they are in low concentration. Once evenly distributed the *net* movement of molecules stops.

Factors which affect the rate of diffusion

* Surface area.
  the greater the surface area the greater the rate of diffusion
* The difference in concentration on either side of the membrane.
  the greater the difference the greater the rate
* The size of molecules.
  smaller molecules may pass through the membrane faster than larger ones
* The presence of pores in the membrane.
  pores can speed up diffusion
* The width of the membrane.
  the thinner the membrane the faster the rate.

Facilitated diffusion

This is a special form of diffusion in which protein carrier molecules are involved. It is much faster than regular diffusion because of the carrier molecules. Each carrier will only bind with a specific molecule. Binding changes the shape of the carrier which then deposits the molecule into the cytoplasm. No energy is used in the process.

Osmosis

This is the movement of water molecules across a selectively permeable membrane:

* from a lower concentrated solution to a higher concentrated solution
* from where water molecules are at a higher concentration to where they are at a lower concentration
* from a hypotonic solution to a hypertonic solution
* from a hyperosmotic solution to a hypo-osmotic solution
* from an area of higher water potential to lower water potential.

> Remember that osmosis is about the movement of **water molecules**. No other substance moves!

The diagram below shows a model of osmosis.

Water molecules move from B to A

> Sometimes the membrane is stated as being selectively permeable, partially permeable or semi-permeable. They all mean the same thing.

selectively permeable membrane

● water molecule
● solute molecule

What is the relationship between water potential of the cell and the concentration of an external solution?

The term 'water potential' is used as a measure of water movement from one place to another in a plant. It is measured in terms of pressure and the units are either kPa (kilopascals) or MPa (megapascals). Water potential is indicated by the symbol ψ *(pronounced psi)*. The following equation allows us to work out the water 'status' of a plant cell.

ψ (cell) water potential (of cell)	=	ψs solute potential (of ions inside cell)	+	ψp pressure potential (of cell wall)	**K E Y P O I N T**

> Remember that water moves from an area of higher water potential to an area of lower water potential. When a cell at -4 MPa is next to a cell at the less negative value the water moves to the more negative value, i.e. -4 MPa > -6 MPa

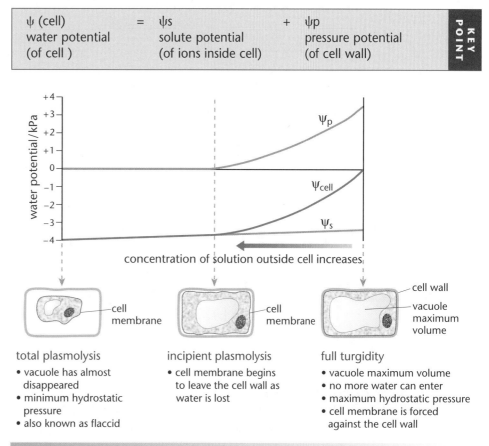

> Note that pressure potential only has a value **above** zero when the cell membrane **begins** to contact the cell wall. The greater the pressure potential the more the cell wall resists water entry. At turgidity $\psi s = \psi p$ when net water movement is zero.

total plasmolysis
- vacuole has almost disappeared
- minimum hydrostatic pressure
- also known as flaccid

incipient plasmolysis
- cell membrane begins to leave the cell wall as water is lost

full turgidity
- vacuole maximum volume
- no more water can enter
- maximum hydrostatic pressure
- cell membrane is forced against the cell wall

Active transport

> Note that active transport is the movement of molecules up a concentration gradient.

In **active transport** molecules move from where they are in lower concentration to where they are in higher concentration. A protein carrier molecule is used. This is **against the concentration gradient** and always **needs energy**. A plant may contain a higher concentration of Mg^{2+} ions than the soil. It obtains a supply by active transport through the cell surface membranes of root hairs. Only Mg^{2+} ions can bind with the specific protein carrier molecules responsible for their entry into the plant. This is also known as active ion uptake, but is a form of active transport. Any process that reduces respiration in cells will reduce active transport, e.g. adding a poison such as cyanide or reducing oxygen availability. This is because energy is needed for the process and this energy is released by respiration.

In the small intestine, glucose is absorbed into the bloodstream indirectly by active transport. Sodium ions are pumped out of the epithelial cells allowing glucose and sodium ions to diffuse into the cell by a co-transporter system.

Endocytosis, exocytosis, pinocytosis and phagocytosis

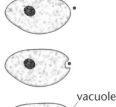

Some substances, often due to their large size, enter cells by **endocytosis** as follows:

- the substance contacts the cell surface membrane which indents
- the substance is surrounded by the membrane, forming a vacuole or vesicle
- each vacuole contains the substance and an outer membrane which has detached from the cell surface membrane.

When fluids enter the cell in this way this is known as **pinocytosis**. When the substances are large solid particles, this is called **phagocytosis**. Some substances leave the cell in a reverse of endocytosis. Here the membrane of the vacuole or vesicle merges with the cell surface membrane depositing its contents into the outside environment of the cell. This is known as **exocytosis**.

Progress check

1. A plant contains a greater concentration of Fe^{2+} ions than the soil in which it is growing. Name and describe the process by which the plant absorbs the ions against the concentration gradient.

2. Explain the following:
   (a) Endocytosis of an antigen by a phagocyte
   (b) Exocytosis of amylase molecules from a cell.

2 (a) **Endocytosis**: antigen contacts the cell membrane of the phagocyte; cell membrane surrounds the antigen, forming a vacuole; the vacuole contains the antigen and an outer membrane which has detached from the cell surface membrane.
(b) **Exocytosis**: a vesicle in the cell contains amylase molecules; the vesicle merges with the cell membrane; amylase contents deposited outside of the cell.

1 **Active transport**: molecules move from a lower concentration to a higher concentration; through the cell surface membranes of root hairs; protein carrier molecules in membranes used; energy needed; Fe^{2+} ions can bind with the protein carrier molecules which allow entry into the plant.

4.3 Gaseous exchange in humans

After studying this section you should be able to:

- *understand why organisms need to be adapted for gaseous exchange*
- *explain how the human breathing system brings about ventilation and gaseous exchange*

LEARNING SUMMARY

How are organisms adapted for efficient gaseous exchange?

EDEXCEL 2.6

The range of respiratory surfaces in this chapter each have common properties, such as high surface area to volume ratio, one cell thick lining tissue, many capillaries.

The exchange of substances across cell surface membranes has been described. Larger organisms have a major problem in exchange because of their low surface area to volume ratio. Some organisms, like flatworms, have a large surface area due to the shape of their body. Other organisms satisfy their needs by having tissues and organs which have special adaptations for efficient exchange. In simple terms, these structures achieve a very high surface area, e.g. a leaf, and link to the transport system to allow import and export from the organ.

Gaseous exchange in humans

EDEXCEL 2.6

The diagram below shows the **human gas exchange system**. The alveoli are the site of gaseous exchange and they are connected to the outside air via a system of branching tubes.

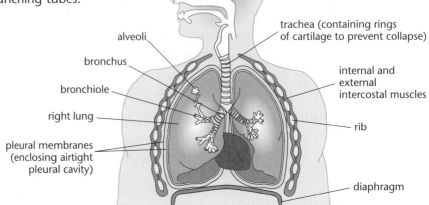

- trachea (containing rings of cartilage to prevent collapse)
- alveoli
- bronchus
- bronchiole
- right lung
- pleural membranes (enclosing airtight pleural cavity)
- internal and external intercostal muscles
- rib
- diaphragm

Ventilation

Drawing air in and out of the lungs involves changes in pressure and volume in the chest. These changes work because the **pleural** membranes form an airtight **pleural cavity**.

Breathing in (inhaling):

1. The external intercostal muscles contract, moving the ribs upwards and outwards.
2. The diaphragm contracts and flattens.
3. Both of these actions will increase the volume in the pleural cavity and so decrease the pressure.
4. Air is therefore drawn into the lungs.

Breathing out (exhaling):

1. The internal intercostal muscles relax and the ribs move down and inwards.
2. The diaphragm relaxes and domes upwards.
3. The volume in the pleural cavity is decreased so the pressure is increased.
4. Air is forced out of the lungs.

Lungs of a mammal

The ventilation mechanism allows inhalation of air, which then diffuses into alveoli to exchange the respiratory gases. Completion of ventilation takes place when gases are expelled into the atmosphere. The diagram on the left shows the structure of the alveoli.

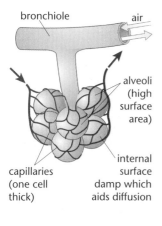

bronchiole

air

alveoli (high surface area)

internal surface damp which aids diffusion

capillaries (one cell thick)

Adaptations of lungs for gaseous exchange

- Air flows through a trachea (windpipe) supported by cartilage.
- It reaches the alveoli via tubes known as bronchi and bronchioles.
- Lungs have many alveoli (air sacs) which have a high surface area.
- Each alveolus is very thin (diffusion is faster over a short distance).
- Each alveolus has an inner film of moisture containing a chemical called surfactant. This reduces the surface tension and makes it easier to inflate the lungs.
- Each alveolus has many capillaries, each one cell thick, to aid diffusion.
- There are many blood vessels in the lungs to give a high surface area for gaseous exchange and transport of respiratory substances.

Measuring ventilation

The process of ventilation can be investigated using a device called a **spirometer**. It can measure the volume of air exchanged in a single breath. This is called the **tidal volume**. It can also measure the number of breaths per minute – the **breathing rate**. If these two figures are multiplied together, the result gives the volume of gas exchanged in one minute, the **pulmonary ventilation**.

pulmonary ventilation = tidal volume × breathing rate

The **maximum** volume of air that can be breathed out in one breath is called the vital capacity.

Sample question and model answer

The diagram below shows two adjacent plant cells A and B.
The water potential of a cell is ψ(cell).

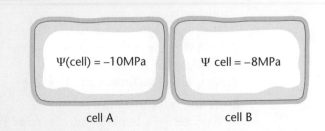

cell A   cell B

(a) (i) Draw an arrow on the diagram to show the direction of water flow. Explain how you worked out the direction. [2]

The arrow should be drawn from cell B to cell A.
Direction from −8 MPa to −10 MPa from a larger to a smaller ψ(cell).

(ii) Name the condition of the cell when ψ(cell) = 0 [1]

full turgor or fully turgid

(b) Give **one** difference between the following terms: [2]

facilitated diffusion

molecules move down a gradient

active transport

energy is needed for the process

Be careful with this type of question. You may believe that 'up a gradient' could be given for active transport. It is correct, but it's too close to the 'down a gradient' idea for facilitated diffusion.

Go for a completely different idea, as shown.

(c) What effect would the following temperatures have on the active transport of Mg^{2+} ions across a cell surface membrane of a plant cell? Assume the plant is a British native. [4]

(i) 30°C

It is likely that active transport would be efficient because the temperature would be ideal for the Mg^{2+} to bind with a carrier molecule.

(ii) 80°C

process likely not to work;
protein carrier denatured;
Mg^{2+} would not be able to bind.

Practice examination questions

1

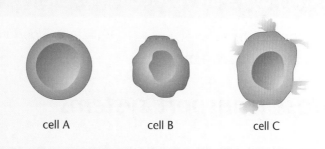

cell A   cell B   cell C

Cells A, B and C have been placed in different concentrations of salt solution.

(a) Explain each of the following in terms of water potential.

   (i) Cell A did not change size at all.

   (ii) Cell B decreased in volume.

   (iii) Cell C became swollen and burst.   [3]

(b) Which process is responsible for the changes to cells B and C?   [1]

2 (a) Give one similarity between active transport and facilitated diffusion.   [1]

  (b) Give one difference between active transport and facilitated diffusion.   [1]

3 Describe and explain how the alveoli of lungs are adapted to efficient gaseous exchange.   [4]

Chapter 5
Transport

The following topics are covered in this chapter:

- Mass transport systems
- Heart: structure and function
- Blood vessels

5.1 Mass transport systems

After studying this section you should be able to:

- explain why most multicellular organisms need a mass transport system
- understand the importance of a high surface area to volume ratio
- describe the differences between open and closed circulatory systems
- understand the implications of using single or double circulatory systems

LEARNING SUMMARY

Why do most multicellular organisms need a mass transport system?

EDEXCEL   1.1.7

The bigger an organism is, the lower its surface area to volume ratio. Substances needed by a large organism could not be supplied through its exposed external surface. Oxygen passing through an external surface would be rapidly used up before reaching the many layers of underlying cells. Similarly waste substances would not be excreted quickly enough. This problem has been solved, through evolution, by specially adapted tissues and organs.

> Leaves, roots, gills and lungs all have high surface area to volume properties so that supplies of substances vital to **all** the living cells are made available by these structures. Movement of substances to and from these structures is carried out by efficient **mass transport systems**.
>
> KEY POINT

This is a little like transport on trains where people travel together on the same train, in the same direction, at the same speed, but may get off at different places.

In a mass transport system, all the substances move in the same direction at the same speed. Across the range of multicellular organisms found in the living world are a number of mass transport systems, e.g. the mammalian circulatory system and the vascular system of a plant.

Mass transport systems are just as important for the rapid removal of waste as they are for supplies. Supplies include an immense number of substances, e.g. glucose, oxygen and ions. Even communication from one cell to another can take place via a mass transport system, e.g. hormones in a blood stream.

The greater the metabolic rate of an organism, the greater the demands on its mass transport system.

5.2 Heart: structure and function

After studying this section you should be able to:

- *recall the structure, cardiac cycle and electrical stimulation of a mammalian heart*

LEARNING SUMMARY

The mammalian heart

EDEXCEL   1.1.7

The heart consists of a range of tissues. The most important one is cardiac muscle. The cells have the ability to contract and relax through the complete life of the person, without ever becoming fatigued. Each cardiac muscle cell is **myogenic**. This means it has its own inherent rhythm. Below are diagrams of the heart and its position in the circulatory system.

Note that tricuspid and bicuspid valves are known as atrioventricular valves.

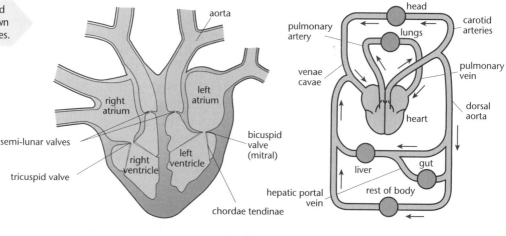

Structure

The heart consists of four chambers, **right** and **left atria** above **right** and **left ventricles**. The functions of each part are as follows.

If blood moved in the wrong direction, then transport of important substances would be impeded.

- The **right atrium** links to the **right ventricle** by the **tricuspid valve**. This valve prevents backflow of the blood into the atrium above, when the ventricle contracts.

- The **left atrium** links to the **left ventricle** by the **bicuspid valve (mitral valve)**. This also prevents backflow of the blood into the atrium above.

- The **chordae tendonae** attach each ventricle to its **atrioventricular valve**. Contractions of the ventricles have a tendency to force these valves up into the atria. Backflow of blood would be dangerous, so the chordae tendonae hold each valve firmly to prevent this from occurring.

Check out these diagrams of a valve.

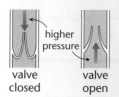

valve closed   valve open

higher pressure

You can work out if a valve is open or closed in terms of pressure. Higher pressure above than below a semi-lunar valve closes it. Higher pressure below the semi-lunar valve than above, opens it.

- Semi-lunar (pocket) valves are found in the blood vessels leaving the heart (pulmonary artery and aorta). They only allow exit of blood from the heart through these vessels following ventricular contractions. Elastic recoil of these arteries and relaxation of the ventricles closes each semi-lunar valve.

- Ventricles have thicker muscular walls than atria. When each atrium contracts it only needs to propel the blood a short distance into each ventricle.

- The left ventricle has even thicker muscular walls than the right ventricle. The left ventricle needs a more powerful contraction to propel blood to the systemic circulation (all of the body apart from the lungs). The right ventricle propels blood to the nearby lungs. The contraction does not need to be so powerful.

Cardiac cycle

Blood must continuously be moved around the body, collecting and supplying vital substances to cells as well as removing waste from them. The heart acts as a pump using a combination of **systole** (contractions) and **diastole** (relaxation) of the chambers. The cycle takes place in the following sequence.

Stage 1

Ventricular diastole, atrial systole
Both ventricles relax simultaneously. This results in lower pressure in each ventricle compared to each atrium above. The atrioventricular valves open partially. This is followed by the atria contracting which forces blood through the atrioventricular valves. It also closes the valves in the vena cava and pulmonary vein. This prevents backflow of blood.

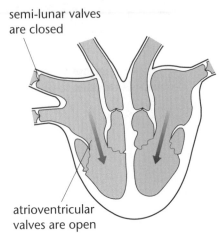

semi-lunar valves are closed

atrioventricular valves are open

Stage 2

Ventricular systole, atrial diastole
Both atria then relax. Both ventricles contract simultaneously. This results in higher pressure in the ventricles compared to the atria above. The difference in pressure closes each atrioventricular valve. This prevents backflow of blood into each atrium. Higher pressure in the ventricles compared to the aorta and pulmonary artery opens the semi-lunar valves and blood is ejected into these arteries. So blood flows through the systemic circulatory system via the aorta and vena cava and through the lungs via the pulmonary vessels.

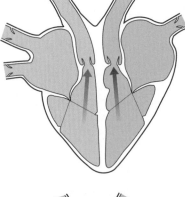

Stage 3

Ventricular diastole, atrial diastole
Immediately following ventricular systole, both ventricles and atria relax for a short time. Higher pressure in the aorta and pulmonary artery than in the ventricles closes the semi-lunar valves. This prevents the backflow of blood. Higher pressure in the vena cava and pulmonary vein than in the atria results in the refilling of the atria.

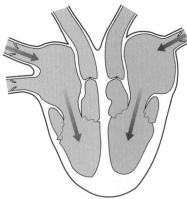

The cycle is now complete – *GO BACK TO STAGE 1!*

The whole sequence above is **one** cardiac cycle or heartbeat and it takes less than one second. The number of heartbeats per minute varies to suit the activity of an organism. Vigorous exercise is accompanied by an increase in heart rate to allow faster collection, supply and removal of substances because of enhanced blood flow. Conversely during sleep, at minimum metabolic rate, heart rate is correspondingly low because of minimum requirements by the cells.

Examination questions often test your knowledge of the opening and closing of valves. Always analyse the different pressures given in the question. A greater pressure behind a valve opens it. A greater pressure in front closes it.

Returning to Stage 1, the cycle begins again. The hormone adrenaline increases the heart rate still further. Even your examinations may increase your heart rate!

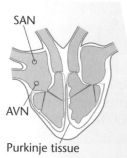

SAN

AVN

Purkinje tissue

The SAN is the natural pacemaker of the heart.

All of the Purkinje fibres together are known as the **Bundle of His**.

This is one of the examiners' favourite ways to test heart-related concepts. Look at the **peak** of the **ventricular contraction**. It coincides with the **trough** in the **ventricular volume**. This is not surprising, because as the ventricle contracts it empties! Use the data of higher pressure in one part and lower in another to explain:

(a) movement of blood from one area to another
(b) the closing of valves.

How is the heart rate controlled?

It has already been stated that the cardiac muscle cells have their own inherent rhythm. Even an individual cardiac muscle cell will contract and relax on a microscope slide under suitable conditions. An orchestra would not be able to play music in a coordinated way without a conductor. The cardiac muscle cells must be similarly coordinated, by a **pacemaker** area in the heart. Electrical stimulation from the brain can alter the activity of the pacemaker and therefore change the rate and strength of the heartbeat.

* The heart control centre in the brain is in the medulla oblongata.
* The sympathetic nerve stimulates an increase in heart rate.
* The vagus nerve stimulates a decrease in heart rate.
* These nerves link to a pacemaker structure in the wall of the right atrium, the **sinoatrial node (SAN)**.
* A wave of electrical excitation moves across both atria.
* They respond by contracting (the right one slightly before the left).
* The wave of electrical activity reaches a second pacemaker, the **atrioventricular node (AVN)**, which conducts the electrical activity through the **Purkinje fibres**.
* These Purkinje fibres pass through the septum of the heart deep into the walls of the left and right ventricles.
* The ventricle walls begin to contract from the apex (base) upwards.
* This ensures that blood is ejected efficiently from the ventricles.

Graphs to show the changes in pressure and volume during the cardiac cycle

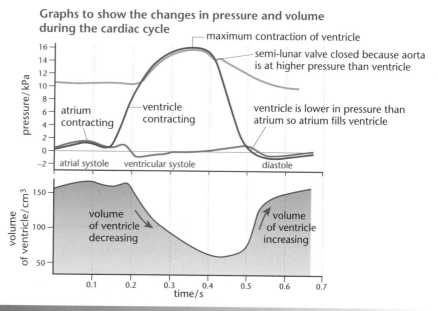

Progress check

The medulla oblongata can increase the heart rate. The statements below include all of the events which take place, but in the wrong order. Write them out in the correct sequence.

A this ensures that blood is ejected efficiently from the ventricles

B the wave of electrical activity reaches the **atrioventricular node (AVN)** which conducts the electrical activity through the **Purkinje fibres**

C a wave of electrical excitation moves across both atria

D the sympathetic nerve conducts electrical impulses

E electrical impulses are received at the **sinoatrial node (SAN)**

F as a result the atria contract

G the ventricle walls begin to contract from the apex (base) upwards

D E C F B G A

5.3 Blood vessels

After studying this section you should be able to:

- describe the structure and functions of arteries, capillaries and veins
- understand the importance of valves in the return of blood to the heart
- understand the difference between plasma, tissue fluid and lymph

LEARNING SUMMARY

Arteries, veins and capillaries

EDEXCEL ▶ 1.1.8

The blood is transported to the tissues via the vessels. The main propulsion is by the ventricular contractions. Blood leaves the heart via the arteries, reaches the tissues via the capillaries, then returns to the heart via the veins. Each blood vessel has a space through which the blood passes; this is the **lumen**. The structure of the vessels is shown below.

Artery

Note that the pressure in the **arteries** is highest because:

(a) they are closest to the ventricles
(b) they contract forcefully themselves.

Capillaries are the next highest in pressure, the main factor being their resistance to blood flow.

Finally, the pressure of **veins** is the lowest because:

(a) they are furthest from the ventricles
(b) they have a low amount of muscle.

If given blood pressures of vessels, be ready to predict the correct direction of blood flow.

- It has a thick **tunica externa** which is an outer covering of tough collagen fibres.
- It has a **tunica media** which is a middle layer of **smooth muscle** and **elastic fibres**.
- It has a lining of **squamous endothelium** (very thin cells).
- It can contract using its **thick muscular layer**.

tunica (collagen externa fibres)
lumen
tunica media
endothelial lining

Capillary

- It is a very thin blood vessel, the endothelium is just one cell thick.
- Substances can exchange easily.
- It has such a high resistance to blood flow that blood is slowed down. This gives more time for efficient exchange of chemicals at the tissues.

endothelium
lumen

Vein

- It has a thin **tunica externa** which is an outer covering of tough collagen fibres.
- It has a very thin **tunica media** which is a middle layer of **smooth muscle** and **elastic fibres**.
- It has a lining of **squamous endothelium** (very thin cells).
- It is lined with semi-lunar valves which prevent the backflow of blood.

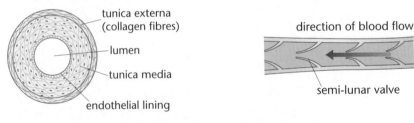

tunica externa
(collagen fibres)
lumen
tunica media
endothelial lining

direction of blood flow
semi-lunar valve

How do the veins return the blood to the heart?

Veins have a thin tunica media, so only mild contractions are possible. They return blood in an unexpected way. Every time the organism moves physically, blood is squeezed between skeletal muscles and forced along the vein.

Blood must travel towards the heart because of the direction of the semi-lunar valves. Any attempt at backflow and the semi-lunar valves shut tightly!

Capillary network

Every living cell needs to be close to a **capillary**. The arteries transport blood from the heart but before entry into the capillaries it needs to pass through smaller vessels called **arterioles**. Many arterioles contain a ring of muscle known as a **pre-capillary sphincter**. When this is contracted the constriction shuts off blood flow to the capillaries, but when it is dilated, blood passes through. Some capillary networks have a shunt vessel. When the sphincter is constricted blood is diverted along the shunt vessel so the capillary network is by-passed. After the capillary network has permeated an organ the capillaries link into a **venule** which joins a **vein**.

In the skin the superficial capillaries have the arteriole/shunt vessel/venule arrangement as shown opposite. When the arteriole is dilated (**vasodilation**) more heat can be lost from the skin. When the arteriole is constricted (**vasoconstriction**) the blood cannot enter the capillary network so is diverted to the core of the body. Less heat is lost from the skin.

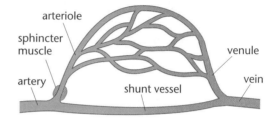

Although the pressure of the blood in the capillaries is lower than in the arteries or arterioles, there is still enough pressure to force out some of the liquid part of the blood. The liquid part of the blood is called **plasma** and when it is forced out of the capillaries it is called **tissue fluid**.

This tissue fluid bathes the cells, supplying them with nutrients and taking up waste products. At the venous end of the capillary bed, most of this tissue fluid is reabsorbed back into the capillaries.

Lymphatic system

There is a network of vessels other than the blood system. They are the **lymphatic vessels**. They collect any tissue fluid that is not reabsorbed back into the capillaries.

The lymph vessels have valves to ensure transport is in one direction. Along some parts of the lymphatic system are lymph nodes. These are swellings lined with white blood cells (macrophages and lymphocyte cells). The lymph fluid is finally emptied back into the blood near the heart.

Progress check

The diagram shows the structure of a blood vessel.

(a) (i) Which type of vessel, artery, vein or capillary is shown? Give a reason for your choice.

   (ii) What is the function of the tunica media?

(b) The pressure values 30 kPa, 10 kPa and 5 kPa correspond to the different types of vessel. Give the correct value for each vessel so that blood flows around the body.

tunica (collagen externa fibres)

lumen

tunica media

endothelial lining

(b) artery, 30 kPa, capillary, 10 kPa and vein, 5 kPa.
(ii) contracts to help transport blood.
(a) (i) artery; the vessel has a thick tunica externa

Sample question and model answer

The diagram shows **one** stage in the cardiac cycle.

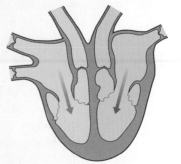

Always look for the valves. If the heart valve is open then the chamber behind it is contracting.

(a) Which stage of the cardiac cycle is shown in the diagram? Give **two** reasons for your answer. [3]

atrial systole
the atrioventricular valves are open/blood flows through the atrioventricular valves,
semi-lunar valves are closed.

(b) Write an X in one chamber to show the position of the atrioventricular node (AVN). [1]

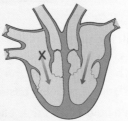

(c) How does the AVN stimulate the contraction of the ventricles? [1]

Passes electrical impulses to Purkinje tissue/Bundle of His.

Practice examination questions

1 The diagram shows a capillary bed in the upper part of the skin. The arteriole is constricted.

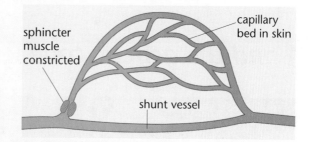

Use the information in the diagram and your own knowledge to answer the questions below.

(a) As a result of arteriole constriction, to where would the blood flow? [1]

(b) Explain how this would help maintain the body temperature. [4]

2 The diagram shows the pressure of the blood as it passes through different parts of the circulatory system.

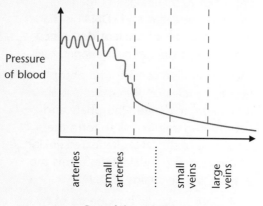

(a) Which blood vessel is missing from the diagram? [1]

(b) How does the blood flow in an artery differ to the flow of blood in a vein? [2]

(c) Describe how veins are adapted to ensure that blood is able to return to the heart [3]

Chapter 6

Genes and cell division

The following topics are covered in this chapter:

- DNA and the genetic code
- Cell division
- Gene technology

6.1 DNA and the genetic code

After studying this section you should be able to:

- *recall the structure of DNA*
- *outline the roles of DNA and RNA in the synthesis of protein*
- *use organic base codes of DNA and RNA to identify amino acid sequences*

LEARNING SUMMARY

Deoxyribonucleic acid (DNA) and chromosome structure

EDEXCEL   1.2.10

Each chromosome in a nucleus consists of a series of genes. A gene is a section of a chemical called **DNA** and each gene controls the production of a polypeptide important to the life of an organism. **Deoxyribonucleic acid (DNA)** is made up of a number of **nucleotides** joined together in a double helix shape.

> You need to be aware that many nucleotides join together to form the polymer, DNA.

Each nucleotide consists of a phosphate group, a molecule of deoxyribose sugar and an organic base. Phosphate and pentose sugar units link to form the backbone of the DNA. Repeated linking of the monomer nucleotides forms the polynucleotide chains of DNA.

> Each strand of DNA is said to be complementary to the other. **Examination tip:** be ready to identify one strand when given the matching complementary strand.

Each DNA molecule is made up of two polynucleotide chains. The two chains are held together by hydrogen bonds between the bases.

The organic base of each nucleotide can be any one of **adenine, thymine, cytosine** or **guanine**. Adenine forms hydrogen bonds with thymine and cytosine with guanine.

The two chains then twist up to form a double helix.

Why does the DNA of one organism differ from the DNA of another?

Differences in the DNA of organisms such as humans and houseflies lie in the **different sequences** of the organic bases. Each sequence of bases is a code to make a protein, usually vital to the life of an organism.

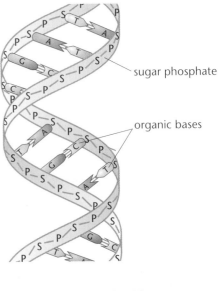

sugar phosphate

organic bases

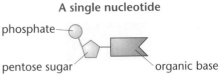

A single nucleotide

phosphate

pentose sugar   organic base

DNA

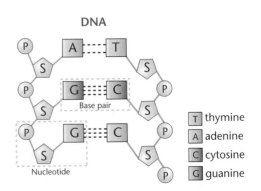

Base pair

Nucleotide

T	thymine
A	adenine
C	cytosine
G	guanine

How does DNA control protein production?

EDEXCEL ▶ 1.2.12-14

Each different polypeptide is made of a specific order of amino acids and so DNA must code for this order.

> **KEY POINT**
> Each three adjacent bases code for one amino acid in the polypeptide. The genetic code is therefore a **triplet** code.

The table below shows all the triplet sequences of organic bases found along DNA strands and the coding function of each.

Use the key to identify the amino acids in the table opposite.

Amino acid	Abbreviation
alanine	Ala
arginine	Arg
asparagine	Asn
aspartic acid	Asp
cysteine	Cys
glutamine	Gln
glutamic acid	Glu
glycine	Gly
histidine	His
isoleucine	Iso
leucine	Leu
lysine	Lys
methionine	Met
phenylalanine	Phe
proline	Pro
serine	Ser
threonine	Thr
tryptophan	Trp
tyrosine	Tyr
valine	Val

Genetic code functions of DNA

		Second organic base							Third organic base
		A		**G**		**T**		**C**	
A		A A A	Phe	A G A		A T A	Tyr	A C A	A
		A A G		A G G	Ser	A T G		A C G	Cys / G
		A A T	Leu	A G T		A T T	stop	A C T	stop / T
		A A C		A G C		A T C	stop	A C C	Trp / C
G		G A A		G G A		G T A	His	G C A	A
		G A G	Leu	G G G	Pro	G T G		G C G	Arg / G
		G A T		G G T		G T T	Gln	G C T	T
		G A C		G G C		G T C		G C C	C
T		T A A		T G A		T T A	Asn	T C A	Ser / A
		T A G	Ile	T G G	Thr	T T G		T C G	G
		T A T		T G T		T T T	Lys	T C T	Arg / T
		T A C	Met	T G C		T T C		T C C	C
C		C A A		C G A		C T A	Asp	C C A	A
		C A G	Val	C G G	Ala	C T G		C C G	Gly / G
		C A T		C G T		C T T	Glu	C CT	T
		C A C		C G C		C T C		C C C	C

Do not learn all of the triplet codes. Be ready to use the supplied data in the examination. You will be given a key of different codes and functions.

If you are given a table of codes check them carefully. If the bases are from mRNA then there will be uracil in the table.

Each triplet code is **non-overlapping**. This means that each triplet of three bases is a code, then the next three, and so on along the DNA.

- AAA codes for the amino acid phenylalanine
- GAG codes for the amino acid leucine
- GAC codes for the amino acid leucine

There are more triplet codes than there are amino acids. This is known as the **degenerate code**, because an amino acid such as leucine can be coded for by up to six different codes. Some triplets do not code for amino acids but mark the beginning or end of polypeptides. They are **stop or start** triplets.

RNA and protein synthesis

DNA is found in the nucleus but proteins are made on ribosomes in the cytoplasm. Therefore a messenger is needed to transfer the code. This messenger is a molecule called ribonucleic acid (**RNA**).

RNA is a nucleic acid, made up of nucleotides like DNA but it has some important differences:

DNA	RNA
Two polynucleotide strands	One polynucleotide strand
Contains adenine, cytosine, guanine and thymine	Contains adenine, cytosine, guanine and the base uracil instead of thymine
Contains deoxyribose sugar	Contains ribose sugar

Messenger RNA is formed in the nucleus by making a complementary copy of the DNA coding for the polypeptide. This is called **transcription**.

Here is an example of a coding strand of DNA:

DNA   A A A G A G G A C A C T   *(coding strand)*
mRNA   U U U C U C C U G U G A   *(messenger RNA)*

guanine (G) on DNA codes for cytosine (C) on mRNA

cytosine (C) on DNA codes for guanine (G) on mRNA

thymine (T) on DNA codes for adenine (A) on mRNA

adenine (A) on DNA codes for uracil (U) on mRNA

At the **ribosomes** the message is converted into an amino acid sequence using tRNA. This is called **translation**.

> This type of RNA is called messenger RNA (mRNA) because it carries the message out of the nucleus. There are two other types of RNA called rRNA and tRNA.

Progress check

(a) Name the parts of a nucleotide.
(b) (i) By which bonds do the two strands of DNA link together?
     (ii) How would these bonds be broken in the laboratory to produce single strands of the DNA?
(c) Which organic base is found in DNA but not in RNA?

(c) thymine
(b) (i) hydrogen bonds (ii) heat
(a) pentose sugar, phosphate and organic base. The organic base may be thymine, adenine, cytosine or guanine

6.2 Cell division

After studying this section you should be able to:

- *describe and explain the semi-conservative replication of DNA*
- *understand that DNA must replicate before cell division can begin*
- *recall the purpose of mitosis and meiosis*
- *recall the process of fertilisation in mammals and flowering plants*
- *recognise each stage of the cell cycle and cell division by mitosis*

LEARNING SUMMARY

How do cells prepare for division?

EDEXCEL   1.2.11

Before cells divide they must first make an exact copy of their DNA by using a supply of organic bases, pentose sugar molecules and phosphates. The method by which DNA is copied is called **semi-conservative replication**. The diagram (right) shows this taking place.

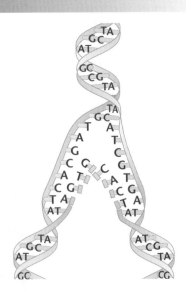

- The DNA begins to unwind under the influence of the enzyme DNA helicase.
- Hydrogen bonds between the two chains then break and the two strands separate.
- Each complementary strand then acts as a template to build its opposite strand from free nucleotides.
- The enzyme DNA polymerase joins the nucleotides together. This process results in the production of two identical copies of double-stranded DNA.

> Remember that as the DNA unwinds each single strand is a **complement** to the other. This means that each has the **matching** series of organic bases.

Evidence for the semi-conservative replication of DNA

Exam questions are often based on experimental data. Apply your knowledge of principles learned during the course and this will be helpful.

The first real evidence came from the results of an experiment carried out by two researchers, Meselson and Stahl.

Bacteria were cultured with a heavy isotope of nitrogen located in the organic bases of their DNA.

The bacteria were then supplied with bases containing the normal light nitrogen atoms. They replicated their DNA using these bases. Their population increased.

Each molecule of DNA of the next generation had one strand containing heavy nitrogen and one strand containing light nitrogen. The mass of the DNA was therefore midway between the original heavy form and normal light DNA.

Semi-conservative replication

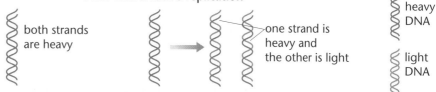

both strands are heavy

one strand is heavy and the other is light

heavy DNA

light DNA

Make sure that you can predict what would happen if the bacteria replicated again.

Semi-conservative therefore means that as DNA splits into its two single strands, each of the new strands is made of newly acquired bases. The other strand, part of the original DNA, remains.

Types of cell division

EDEXCEL ▷ 1.2.11
2.3.6, 2.3.8

Cells divide for the purposes of growth, repair and reproduction. Not all cells can divide but there are two ways in which division may occur: **mitosis** and **meiosis**.

In most organisms, the chromosomes in each cell can be arranged in pairs. Each cell therefore has two copies of each gene.

When the chromosomes are in pairs, the cell is said to be **diploid** and the pairs are called homologous pairs.

Be careful of the spelling of mitosis and meiosis; examiners may penalise you if they are spelt incorrectly.

Mitosis produces genetically identical copies of cells with the same number of chromosomes for growth and repair.

Meiosis produces cells that have half the number of chromosomes, one from each pair. These cells are **haploid** and are used as gametes.

Meiosis introduces variation because the pairs of chromosomes can be split up in many different ways. This is called independent assortment.

Fertilisation

EDEXCEL ▷ 2.3.10

The gametes produced by meiosis can join together in fertilisation. This restores the diploid number of chromosomes again.

The process of fertilisation differs in mammals and flowering plants.

Fertilisation of flowering plants

Pollen reaches a stigma and sticks to the surface, the following events then take place.

- A **pollen tube** begins to grow from the pollen grain.
- The proteins for this tube are produced with the help of the **tube nucleus** which remains near the tip of the tube.
- The **generative nucleus** divides into **2 male nuclei** which follow down the tube.
- The pollen tube grows into an ovule.
- One male nucleus fuses with the egg cell forming a diploid **zygote**; this divides many times to form the **embryo** of the seed.
- The other male nucleus fuses with the two polar nuclei forming a triploid cell; this divides many times to form the endosperm (food store) of the mature seed.

- This is double fertilisation because of the fusion to form the **diploid zygote** and the triploid **primary endosperm cell**

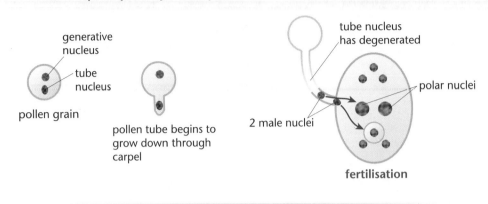

- generative nucleus
- tube nucleus
- pollen grain
- pollen tube begins to grow down through carpel
- 2 male nuclei
- tube nucleus has degenerated
- polar nuclei

fertilisation

Fertilisation in mammals

During copulation the erect penis is inserted into the vagina. When ejaculation takes place the semen is released at the cervix. The semen immediately coagulates so that the female retains the sperm. Within a few minutes the sperm then begin to swim in a fluid produced by the female. Fertilisation normally takes place in the fallopian tube, where one sperm fuses with an ovum. The diagram below shows fertilisation.

> Remember that the ovum can also be called the secondary oocyte

Zona pellucida

The acrosome digests the outer zona pellucida. The head is engulfed so the male nucleus enters.

- Many sperm are attracted to the ovum by chemotaxis.
- The acrosome, a vesicle containing hydrolytic enzymes, breaks down part of the **zona pellucida** to allow the sperm entry into the secondary oocyte.
- The cell membrane of the sperm head fuses with the cell membrane of the secondary oocyte.
- The sperm nucleus is engulfed and moves into the cytoplasm of the secondary oocyte.
- Finally the two sets of haploid nuclei fuse together to form the diploid nucleus of the zygote, which may go on to produce the fetus.

The cell cycle and mitosis

EDEXCEL 2.3.6, 2.3.7

The length of time between a cell being formed and it dividing is called the **cell cycle**. This can be divided up into a number of different phases:

- G1 phase – the cell grows making new proteins and more organelles
- S phase – the DNA of the chromosomes is replicated by semi-conservative replication
- G2 – more organelles are made and a spindle forms
- M phase – this is mitosis involving the separating of the genetic material into two nuclei
- C phase – cytokinesis, where the cell divides into two.

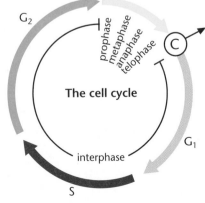

G_2, prophase, metaphase, anaphase, telophase, C, The cell cycle, G_1, interphase, S

Remember that DNA replication takes place before cell division in **interphase** (see page 58). This is not an integral phase of mitosis or meiosis.

The length of the cell cycle varies between different types of cells. Some cells never divide once formed but cells of the bone marrow divide about every eight hours. Interphase usually takes up 95% or more of the whole cell cycle.

Be ready to analyse photomicrographs of all phases of mitosis. If you can spot 10 pairs of chromosomes at the end of telophase, then this is the original diploid number of the parent cell.

Interphase is the period of time between cell divisions. It is made up of G1, S and G2 phases. Once cells start mitosis, they all go through a similar sequence of events. This is shown in the diagrams.

1 Prophase

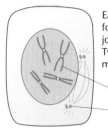

Each chromosome forms two chromatids joined by a centromere. Two centrioles begin to move forming a spindle.

— chromatid

— centriole

2 Metaphase

The chromatids, still joined by a centromere move to the middle of the cell. Each of the two chromatids has identical DNA to the other.

spindle —

3 Anaphase

The spindle fibres join to the centromeres. The spindle fibres shorten and the centromeres split. The separated chromatids are now chromosomes.

4 Telophase

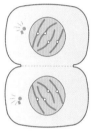

Identical chromosomes move to each pole. The nuclear membrane re-forms. The cell membrane narrows at the middle and two daughter cells are formed.

6.3 Gene technology

After studying this section you should be able to:

- *define different types of stem cells*
- *understand the process of electrophoresis and recall its applications*

LEARNING SUMMARY

Stem cells

EDEXCEL 2.3.11

When cells have differentiated and become specialised they lose their ability to divide to form other types of cells. They also can only divide a limited number of times. **Stem cells** are undifferentiated cells that can divide to form different types of cells. There are different types of stem cells:

- Embryonic stem cells are **totipotent**. This means that they can form any type of cell.
- Later in the embryo and in the adult, the stem cells are **pluripotent**. This means that they can form certain types of cells.

Stem cells have the potential to be used in many types of medical therapies but the use of embryonic stem cells, in particular, has raised a number of ethical issues.

Sample questions and model answers

The table below shows some mRNA codons and the amino acids which are coded by them.

	second position				
	U	C	A	G	
first position U	Phe	Ser	Tyr	Cys	U
	Phe	Ser	Tyr	Cys	C
	Leu	Ser	stop	stop	A
	Leu	Ser	stop	Trp	G

third position

Key to amino acids

Ser – serine   Tyr – tyrosine
Phe – phenylalanine   Trp – tryptophan
Leu – leucine   Cys – cysteine

Use the information in the table to help you answer the following questions.

1

(a) Give a sequence of mRNA bases which would code for leucine.   [1]

UUA or UUG

(b) What does the mRNA base sequence UAC code for?   [1]

Tyrosine

2

The mRNA sequence UCA codes for serine. Work out the base pairs on the DNA. [3]

UCA is coded for by these bases: AGT
AGT links to the bases TCA
So the DNA is AGT
           TCA

3

Use evidence from the table to show that serine is an example of the degenerate code.   [1]

It is coded for by four different base sequences.

4

UAG codes for 'stop'. Explain the effect of the 'stop' code during the process of protein synthesis.   [2]

It is responsible for the polypeptide being terminated which allows it to leave the ribosome once all the amino acids have been linked.

Practice examination questions

1 The diagram below shows a stage in the process of mitosis.

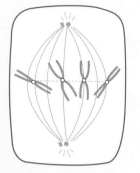

(a) Give the stage of mitosis shown. [1]

(b) How many chromosomes would there be in the daughter cells? [2]

2 The table below shows the relative organic base proportions found in human, sheep, salmon and wheat DNA.

Organism	Proportion of organic bases in DNA (%)			
	Adenine	Guanine	Thymine	Cytosine
human	30.9	19.9	29.4	19.8
sheep	29.3	21.4	28.3	21.0
salmon	29.7	20.8	29.1	20.4
wheat	27.3	22.7	27.1	22.8

(a) Refer to the proportion of organic bases in salmon DNA to explain the association between specific bases. [2]

(b) Suggest a reason for the small difference in proportion of the organic bases adenine and thymine in sheep. [1]

(c) All species possess adenine, guanine, thymine and cytosine in their DNA. Account for the fact that each species is different. [2]

Chapter 7

Classification and biodiversity

The following topics are covered in this chapter:

- *Classification*
- *Biodiversity*

7.1 Classification

After studying this section you should be able to:

- *describe some of the more recent techniques used to classify organisms*

Modern classification techniques

EDEXCEL   2.4.16

For centuries, organisms were classified according to observable physical features. This might be the structure of an animal's skull or the shape of a plant's leaves. Scientists can now use a range of different techniques.

- Microscopic structure – modern electron microscopes have shown that bacteria and other single-celled organisms such as amoeba have completely different cell structure and so they are now put into different kingdoms: *Prokaryotae* and *Protoctista*.
- Genetic differences – it is now possible to compare the DNA base sequences of different organisms. This can be done by a process called **DNA hybridisation**. Short, single-stranded sections of DNA are produced from one species and the extent to which they bind with DNA from another species is measured.
- Biochemical differences – the occurrence of different biochemical molecules is often a good indicator of relationships. It is also possible to work out and compare the amino acid sequence of common proteins.
- Immunological evidence – antibodies against human proteins can be made and then tested against proteins from other animals. The more effective the antibodies are, then the closer the evolutionary relationship is between humans and the other animal.

Immunological results give the following % similarities to man:

chimpanzee 97%,
gibbon 92%,
lemur 37%,
pig 8%

The aim of modern classification systems is to use a range of techniques to produce a system that classifies organisms based on their **evolutionary relationships**.

> A system based on evolutionary relationships is called a **phylogenetic system**.
>
> KEY POINT

7.2 Biodiversity

After studying this section you should be able to:

- *understand what is meant by the term biodiversity*
- *explain why there is so much biodiversity on Earth*

What is biodiversity?

EDEXCEL 2.4.13–14

Biodiversity is a measure of the variation between different living organisms. It could be:

- **genetic diversity** – the differences between the genes in a species
- **species diversity** – the number of different species in a community
- **ecosystem or habitat diversity** – the variety of different areas where organisms can live.

A few definitions

A **habitat**: an area in which an organism lives.

A **population**: all the organisms of one species living in a habitat.

A **community**: all the different species (different populations) living in a habitat.

An **ecosystem**: all the living organisms (biotic) and inorganic parts (abiotic) of a habitat.

A **niche** is where an organism lives and its role in that ecosystem.

A pond is a habitat containing populations such as stickleback and pondweed. All the organisms make up the community and the organisms, the water, the mud, etc. are an ecosystem.

Measuring biodiversity

Biodiversity can be measured within one species.

All the genes that are found in a species or in a population of that species is called the **gene pool**.

The amount of variation can be measured by looking at the variety of different alleles in the gene pool.

The numbers of Hawaiian goose dropped to about 30 birds in 1950 due to a number of factors such as over-hunting. Their numbers are starting to recover but their genetic diversity is really low.

Biodiversity can be measured in a habitat by measuring the **species richness**. This is usually done by calculating an **index of biodiversity**.

An index of biodiversity is used as a measure of the range and numbers of species in an area

$$\text{index d} = \frac{N(N-1)}{\Sigma\, n(n-1)}$$

N = total no. of all individuals of all species in the area

n = total no. of individuals of one species in an area

Σ = the sum of

Consider this example of animals in a small pond

crested newt	8
stickleback	20
leech	15
great pond snail	20
dragonfly larva	2
stonefly larva	10
water boatman	6
caddisfly larva	30
	N = 111

In another pond there were:

crested newt	45
stickleback	4
leech	18
great pond snail	10

d = 2.6

Look at both indices. 6.05 is an indicator of greater diversity. The higher number indicates greater diversity.

$$d = \frac{111 \times 110}{(8 \times 7) + (20 \times 19) + (15 \times 14) + (20 \times 19) + (2 \times 1) + (10 \times 9) + (6 \times 5) + (30 \times 29)}$$

$$d = \frac{12\,210}{2018} \qquad \text{so} \quad d = 6.05$$

Sources of biodiversity

EDEXCEL 2.3.14
2.3.15
2.4.14
2.4.15
2.4.17

What is variation?

Species throughout the biosphere differ from each other.

Variation describes the differences which exist in organisms throughout the biosphere. This variation consists of differences **between** species as well as differences **within** the same species. An organism's characteristics (phenotype) is a combination of the action of an organism's genes (genotype) and the influence of the environment.

genotype + environment = phenotype

Therefore variation can depend on environmental factors and genetic diversity. The variation that is produced can be either continuous or discontinuous.

Continuous variation

This is shown when there is a range of **small incremental differences** in a feature of organisms in a population. An example of this is height in humans. If the height of each pupil in a school is measured then from the shortest pupil to the tallest, there are very small differences across the distribution. This is shown by the graph below which shows smooth changes in height across a population. This type of variation is shown when features are controlled by **polygenic inheritance**. A number of genes **interact** to produce the expressed feature.

Remember that a species shows continuous variation when there are small incremental differences, e.g. height of people in a town. Beginning with the smallest and ending with the tallest there would probably be at least one person at each height, at 1cm increments. A smooth gradation of differences!

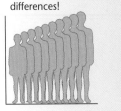

Height of person

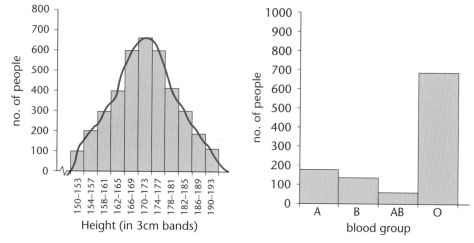

Discontinuous variation

This is shown when a characteristic is expressed in discrete categories. Humans have four discrete blood groups, A, B, AB or O. There are no intermediates, the differences are clear cut!

Why do organisms show variation?

The many different habitats on Earth provide different conditions for organisms to live in. There may be variations in a number of different factors that can affect organisms:

- **climatic** factors, e.g. temperature, water availability, light
- **edaphic** (soil) factors, e.g. pH, minerals and oxygen content
- **biotic** factors, e.g. competition.

Organisms have become adapted to living in a particular habitat with a particular set of factors. These factors may be behavioural, physiological and anatomical. The following table shows some examples.

Type of organism	Adaptation	Behavioural, physiological or anatomical
cacti	leaves are spines to reduce water loss	anatomical
birds	parental care shown to young	behavioural
carnivores and herbivores	particular dentitions adapted to deal with specialised diets	anatomical
llamas	production of haemoglobin molecules with a higher affinity for oxygen	physiological

How is variation produced?

The vast biodiversity of organisms on Earth has been produced by the process of **natural selection** over millions of years. These are the key features of natural selection:

The survival of the best adapted organisms is often called 'survival of the fittest'.

Populations of organisms show variation and some of this variation can be inherited.

More organisms are born than can survive in a particular environment.

There must be a struggle for survival. Those organisms that are best adapted to the environment will survive.

The organisms that survive will breed and pass on their genes.

The theory of natural selection was first put forward by Charles Darwin in 1858 in his book *On the Origin of Species*. The small changes to populations that occur over long periods of time could result in the formation of new species. This is called **speciation**.

Recently, natural selection has been used to explain the development of:

- pesticide resistance in insects
- drug resistance in microorganisms.

Maintaining biodiversity

We have now started to realise the importance of maintaining biodiversity.

Various methods are being used to try and conserve endangered species and their genetic diversity:

The Millennium Seed Bank project aims to have saved seeds from 25% of the worlds plants by 2020

- Captive breeding programmes in zoos
- Setting up seedbanks
- Reintroduction animals and plants back into the wild
- Educating people about the importance of biodiversity.

Sample question and model answer

The quagga is an extinct mammal that looked like a zebra but had stripes only on its head, neck and forebody. The quagga lived in Africa until the last animal was shot in the late 1870s. The last specimen died in 1883 in a zoo.

The quagga, named *Equus quagga*, was originally thought to be a separate species to the plains Zebra.

Over the last fifty years or so, the markings of many individual zebras have been recorded by scientists.

Because of the great variation in stripe patterns, taxonomists were left with a problem. Was the quagga a separate species or was it a type of zebra? The quagga was the first extinct creature to have its DNA studied. Recent research using DNA from preserved specimens has demonstrated that the quagga was not a separate species at all, but a variety of the plains zebra, *Equus burchelli*.

(a) What type of variation does the stripe pattern in zebras demonstrate? [1]

Continuous variation

(b) How could scientists use the DNA to investigate the relationship between the zebra and the quagga? [2]

Use DNA hybridisation.
The DNA is split into single strands and then scientists see how well it binds with zebra DNA strands.

(c) The plains zebra may have more stripes than the quagga because the stripes provide better camouflage on the plains.
How might this pattern have developed by natural selection? [3]

Zebras show variation and are born with different patterns of stripes.
More stripes make the zebra better camouflaged and so more likely to survive.
These zebras survive, breed and pass on the genes for more stripes.

Chapter 8

Human health and disease

The following topics are covered in this chapter:

- Health and lifestyle
- Genetic disorders

8.1 Health and lifestyle

After studying this section you should be able to:

- *understand how coronary heart disease can be affected by lifestyle*

LEARNING SUMMARY

Coronary heart disease and lifestyle

EDEXCEL ▷ 1.1.10–15

Atherosclerosis is a major health problem caused by eating saturated fats. This circulatory disease may develop as follows:

- yellow fatty streaks develop under the lining of the **endothelium** on the inside of an artery
- the streaks develop into a fatty lump called an **atheroma**
- the atheroma is made from **cholesterol** (taken up in the diet as well as being made in the liver)
- dense **fibrous tissue** develops as the atheroma grows
- the endothelial lining can **split**, allowing blood to contact the fibrous atheroma
- the damage may lead to a blood clot and an artery can be blocked.

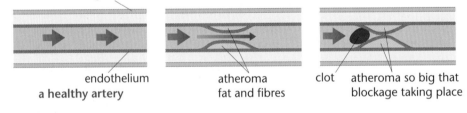

collagen fibres

endothelium
a healthy artery

atheroma
fat and fibres

clot   atheroma so big that
blockage taking place

Remember that the clotting of blood can occur for other reasons. There may be damage at other positions around the body. Blockage of this type is **thrombosis**.

Increasing constriction of an artery caused by **atherosclerosis** and **blood clots** reduces blood flow and increases blood pressure. If the artery wall is considerably weakened then a bulge in the side appears, just like a weakened inner tube on a cycle tyre. There is a danger of bursting and the structure is known as an **aneurysm**.

It is possible for a blood clot formed at an atheroma to break away from its original position. It may completely block a smaller vessel, this is known as an **embolism**.

Coronary heart disease

If the artery which supplies the heart (coronary artery) is partially blocked, then there is a reduction in oxygen and nutrient supply to the heart itself. This is called **coronary heart disease** (CHD). The first sign is often **angina**, the main symptom being sharp chest pains. If total blockage occurs then **myocardial infarction** (heart attack) takes place.

Lifestyle and CHD

Many aspects of lifestyle influence the condition of the cardiovascular system. CHD is a multi-factorial disease and the risk of developing CHD depends on a number of factors.

Statistically people have a greater chance of avoiding CHD if they:

- consume a low amount of saturated fat in their diet
- do not drink alcohol excessively
- consume a low amount of salt
- do not smoke
- are not stressed most of the time
- exercise regularly.

New drugs called statins are being prescribed in an effort to reduce CHD. They are thought to work by decreasing LDL levels in the blood.

Saturated fats (see lipid structure pages 19–20) are found in large quantities in animal tissues. Eating large quantities of saturated fats seems to increase the risk of CHD. It seems that saturated fat increases the levels of **low-density lipoproteins** (LDLs) in the blood. LDLs transport cholesterol in the blood and high levels of LDLs seem to increase the risk of atheroma formation. Other fats such as polyunsaturated fats seem to increase the levels of **high-density lipoproteins** (HDLs). This seems to give some protection against CHD.

Exercise has a **protective effect** on the **heart** and **circulation**. Activities such as jogging, walking, swimming and cycling can:

- reduce the resting heart rate
- increase the strength of contraction of the heart muscle
- increase the stroke volume of the heart (the volume of blood which is propelled during the contractions of the ventricles).

High salt levels have been shown to increase blood pressure, therefore increasing the risk of damage to atheromas.

Smoking also increases blood pressure and makes the blood more likely to clot.

Progress check

(a) Name a specific substance in food which can result in atherosclerosis.
(b) Describe and explain the structural changes which take place in a blood vessel as atherosclerosis develops.
(c) How can the damage caused by an atheroma result in a heart attack?

(a) saturated fat
(b) yellow fatty streaks develop under the cells lining the inside of a blood vessel, the streaks develop into a lump known as an atheroma, the atheroma is made of cholesterol, dense fibrous tissue develops and the lining of the vessel can split.
(c) a blood clot forms which can block the blood vessel completely. Prevention of oxygen supply to the heart results in myocardial infarction (heart attack).

8.2 Genetic disorders

After studying this section you should be able to:

- *Understand the cause of various genetic disorders*
- *Use genetic diagrams to predict the chance of disorders occurring*
- *Explain some possible preventative measures*

Genetic disorders

EDEXCEL   1.2.16

Some diseases, such as CHD, are influenced by environmental factors. Others, such as malaria or cholera, are caused by pathogenic organisms. **Genetic disorders**, however, are caused by our genes and can be passed on from parent to offspring.

Examples of genetic disorders are **cystic fibrosis**, **albinism** and **thalassaemia**.

To understand how these disorders are passed on, we must first understand some principles of genetics.

Essential genetic terms

EDEXCEL   1.2.16

A gene is a section of DNA which controls the production of a protein in an organism. The total effects of all of the genes of an organism are responsible for the characteristics of that organism. Each protein contributes to these characteristics whatever its role, e.g. structural, enzymic or hormonal.

It is necessary to understand the following specialist range of terms used in genetics.

Allele – an alternative form of a gene, always located on the same position along a chromosome.

    E.g. white colour of petals

Dominant allele – if an organism has two different alleles then this is the one which is expressed, often represented by a capital letter.

    E.g. red colour pigment of petals, **R**

Recessive allele – if an organism has two different alleles then this is the one which is **not** expressed, often represented by a lower case letter. Recessive alleles are only expressed when they are not masked by the presence of a dominant allele.

    E.g. white colour pigment of petals, **r**

Check out all of these genetic terms.

- Look carefully at the technique of giving an example with each definition. Often examples help to clarify your answer and are usually accepted by the examiners.
- In examination papers you will need to apply your understanding to new situations.
- Genetics has a specialist language which you will need to use.

Homozygous – refers to the fact that in a diploid organism both alleles are the same.

    E.g. **R R** or **r r**

Heterozygous – refers to the fact that in a diploid organism both alleles are different.

    E.g. **R r** (petal colour would be expressed as red)

Mendel and the laws of inheritance

EDEXCEL   1.2.16

Gregor Mendel was the monk who gave us our understanding of genetics.

He worked with organisms such as pea plants to work out genetic relationships.

> **KEY POINT**
>
> Mendel's first law indicates that:
> * each character of a diploid organism is controlled by a pair of alleles
> * from this pair of alleles only one can be represented in a gamete.

Monohybrid inheritance

> Always show your working out of a genetical relationship in a logical way, just like solving a mathematics problem.

Mendel found that when homozygous pea plants were crossed, a predictable ratio resulted. The cross below shows Mendel's principle.

pea plants   pea plants

T = TALL (dominant)   t = dwarf (recessive)

A homozygous TALL plant was crossed with a homozygous recessive plant

TT   x   tt

gametes   (T)   (T)   (t)   (t)

F1 generation   Tt

All offspring 100% TALL, and heterozygous.

> If you have to choose the symbols to explain genetics, then use something like N and n. Here the upper and lower cases are very different. S and s are corrupted as you write quickly and may be confused by the examiner awarding your marks.

Heterozygous plants were crossed

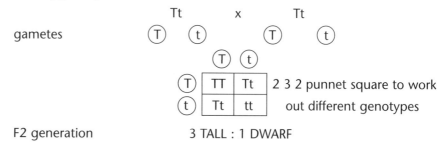

Tt   x   Tt

gametes   (T)   (t)   (T)   (t)

	(T)	(t)
(T)	TT	Tt
(t)	Tt	tt

2 3 2 punnet square to work out different genotypes

F2 generation   3 TALL : 1 DWARF

> In examinations you may have to work out a probability. 3 : 1 is the same as a 1 in 4 chance. Remember only large numbers would confirm the ratio.

In making this cross Mendel investigated one gene only. The height differences of the plants was due to the different alleles. Mendel kept all environmental conditions the same for all seedlings as they developed. The 3:1 ratio of tall to short plants only holds true for large numbers of offspring.

Cystic Fibrosis

EDEXCEL 1.2.16-20

Cystic fibrosis is caused by a recessive allele. During DNA replication, an error can occur and this produced a gene mutation. The resulting allele can be passed on from parent to offspring.

As the allele is recessive, a person only has cystic fibrosis if they have both recessive alleles:

F = the non cystic fibrosis allele
f = the cystic fibrosis allele

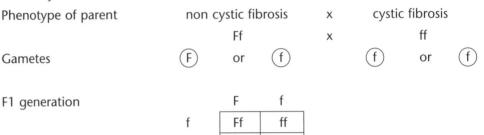

Phenotype of parent	non cystic fibrosis	x	cystic fibrosis
	Ff	x	ff
Gametes	Ⓕ or Ⓕ		Ⓕ or Ⓕ

F1 generation

	F	f
f	Ff	ff
f	Ff	ff

> **Remember** that a ratio is only a probability. If one child with cystic fibrosis is born, the probability of the next one having the disorder is still 1:1.

The ratio of phenotypes that is produced by this cross is therefore

2 non cystic fibrosis (Ff genotypes) : 2 cystic fibrosis (ff genotypes), or 1:1

Symptoms of cystic fibrosis

- In healthy people, a gene codes for a protein which functions as a Cl^- pump in epithelial cells.
- The outward movement of Cl^- ions is accompanied by water, effectively lubricating the outside of these lining cells.
- In people with cystic fibrosis the **Cl^- pump protein** does not function correctly.
- The fluid is **more viscous** and moves in a sluggish movement over the epithelial surfaces.
- Adverse effects are evident in the pancreas-duodenum area where there is **inhibited movement** of substances through the alimentary canal.
- Serious effects are found in the lungs where the epithelial cells secrete **thick sticky mucus**.
- The mucus **inhibits breathing** seriously, so must be removed each day to relieve symptoms.

Genetic screening

There are a number of ways of testing for the cystic fibrosis allele and, therefore, of predicting the chances of this disorder occurring. Testing for the allele is called **genetic screening**. This can be done in a number of ways:

- **Testing individuals** to see if they are carriers of the allele.
- **Pre-implantation genetic testing**, which involves producing an embryo by invitro fertilisation and removing one of the cells for testing.
- **Pre-natal testing**, which involves removing a cell from the foetus for testing. This may be obtained by amniocentesis or chronic villus sampling.

Both of the techniques shown are accompanied by use of an ultrasound scanner. This creates a picture of the fetus in the uterus so that damage can be avoided.

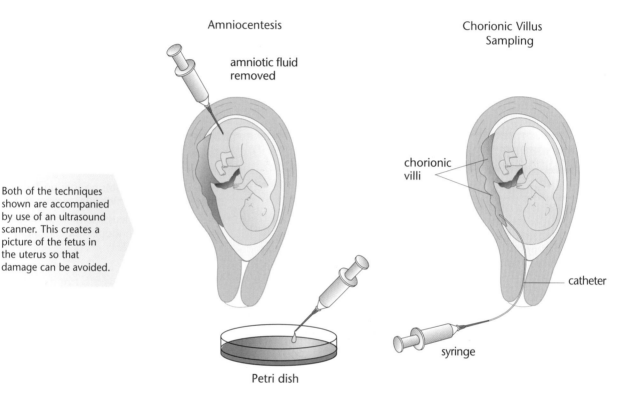

Amniocentesis

amniotic fluid removed

Petri dish

Chorionic Villus Sampling

chorionic villi

catheter

syringe

All of these techniques and the results may produce difficult social and ethical issues that need to be faced.

Gene therapy

Scientists have been trying to devise techniques for changing the defective allele in a person who has cystic fibrosis. This type of technique is called **gene therapy**.

If the alleles are changed in the body cells of a person then this is called **somatic gene therapy**. If the alleles are changed in a gamete then it is called **germ line gene therapy**.

So far, such attempts have achieved limited success.

Practice examination answers

Chapter 1 Biological molecules

1 (a)

RCOOH   HOCH$_2$
|
RCOOH + HOCH
|
RCOOH   HOCH$_2$
fatty acids   glycerol   [2]

(b) Emulsion test: add the sample to ethanol and mix;
decant or pour into water;
if a fat is present a white emulsion forms on the
surface.   [3]
       [Total: 5]

2 (a) peptide bond/peptide link   [1]

(b) –COOH/carboxylic acid   [1]

(c) primary structure; amino acids in a chain   [2]
       [Total: 4]

3 (a) The latent heat of evaporation is large so lots of
energy is needed to evaporate water/in sweating,
much body heat is needed for evaporation.   [2]

(b) High specific heat capacity means that the water
needs a lot of heat energy to increase temperature
significantly, therefore an organism will not
overheat easily.   [2]

(c) Cohesive forces aid the movement of water up the
xylem.   [2]
       [Total: 6]

Chapter 2 Cells

1 (a) A = phospholipid   B = protein   [2]

(b) Acts as a channel/pore; to allow molecules in or
out of the cell.   [1]

(c) One end is hydrophobic and the other is
hydrophilic; Hydrophilic tails 'hide' in the centre of
the membrane;   [2]
       [Total: 5]

2 (a) C = Golgi body
B = Centrioles
A = Cell membrane
E = Mitochondria
D = Rough endoplasmic reticulum   [5]

(b) Correct measure of width nucleus (~2.0 cm)
convert to micrometres (20 000 μm)
divide size by 5 000 (4 μm)   [3]

(c) Liver cells carry out many functions;
Need large amounts of energy/ATP   [2]

       [Total: 10]

3 (a) cell wall (not cellulose); no true nucleus; no
mitochondria; plasmids [any two for 2 marks]   [2]

(b) mitochondria; nucleus; Golgi body; large
ribosomes/rough endoplasmic reticulum
[any two for 2 marks]   [2]
       [Total: 4]

Chapter 3 Enzymes

1 (a) lock and key – the substrate is a similar shape to
the active site; it fits in and binds with the active
site like a key (substrate) fitting into a lock (active
site); induced fit – the substrate is not a matching
'fit' for the active site, but as the substrate
approaches, the active site changes into an
appropriate shape.

(b) reversible – enters active site but will come out again
irreversible – binds permanently with enzymes   [4]
       [Total: 4]

2 The substrate molecule collides with the active site of
the enzyme; as it approaches, the active site changes
shape to become compatible with the substrate
shape.   [2]
       [Total: 2]

Chapter 4 Exchange

1 (a) (i) No change in size because the water potential inside the cell equals the water potential of the solution outside the cell. [1]

   (ii) The water potential of the solution outside the cell is more negative than the water potential inside the cell. [1]

   (iii) The water potential of the solution outside the cell is less negative than the water potential inside the cell. [1]

   (b) osmosis [1]

   [Total: 4]

2 (a) They both use a protein carrier molecule. [1]

   (b) Active transport needs energy or mitochondria, whereas facilitated diffusion does not.
   OR Active transport allows molecules to move from a lower concentrated solution to a higher concentrated solution.
   OR Active transport allows molecules to move against a concentration gradient. [1]

   [Total: 2]

3 Alveoli have a very high surface area; they are very close to many capillaries; capillaries are one cell thick/very thin/have squamous epithelia; they are kept damp which facilitates diffusion. [4]

   [Total: 4]

Chapter 5 Transport

1 (a) to the body core [1]

   (b) less blood reaches the superficial capillaries of the skin; so less heat is lost by conduction, convection and radiation; blood in the body core better insulated by the adipose layer of the skin [4]

   [Total: 5]

2 (a) capillary [1]

   (b) higher pressure;
   fluctuations in pressure; [2]

   (c) valves to prevent backflow;
   wide lumen;
   muscle in wall can contract [3]

   [Total: 6]

Chapter 6 Genes and cell division

1 (a) metaphase [1]

   (b) 4 [2]

   [Total: 3]

2 (a) adenine and thymine are similar proportions because adenine binds with thymine;
   cytosine and guanine are similar proportions because cytosine binds with guanine [2]

   (b) They should be identical in number but the scientists were operating at the limits of instrumentation. [1]

   (c) Organic bases form the codes for different amino acids. Different sequences of amino acids form the different proteins specific to a species. [2]

   [Total: 5]

Notes

Notes

Notes

Index

Index

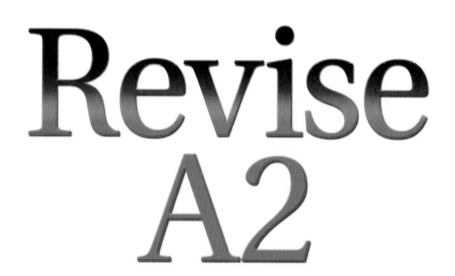

Revise A2

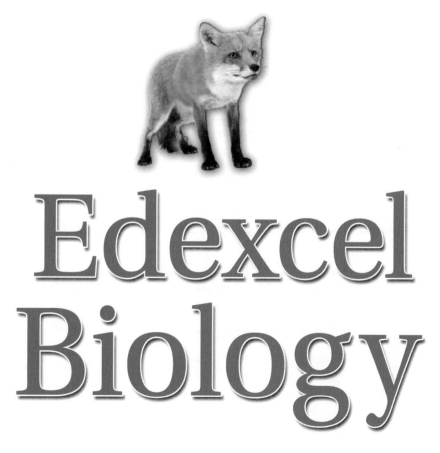

Edexcel Biology

John Parker & Ian Honeysett

Contents

Contents

Specification list

The specification labels on each page refer directly to the units in the exam specification, i.e. EDEXCEL 5.7.2 refers to unit 5, topic 7, item 2.

Edexcel Biology

UNIT	SPECIFICATION TOPIC	CHAPTER REFERENCE	STUDIED IN CLASS	REVISED	PRACTICE QUESTIONS
Unit 4	*Photosynthesis*	*1.1, 1.2*			
	Ecology	*5.1, 5.2, 5.3, 5.4*			
	Global warming	*5.5*			
	Speciation	*4.3*			
	Protein synthesis	*4.1*			
	Gene expression	*4.1*			
	Gene technology	*4.2*			
	Infection and immunity	*6.1, 6.2, 6.3*			
Unit 5	*ATP*	*1.1*			
	Respiration	*1.3*			
	Muscles	*2.4, 2.5*			
	Homeostasis	*3.1, 3.2, 3.3*			
	Nerves	*2.1, 2.4*			
	The brain	*2.3*			
	Coordination in plants	*2.5*			
	Vision	*2.2*			
	Exercise and health	*6.4*			

Examination analysis

Unit 4
This unit is assessed by means of a written examination paper, which lasts 1 hour 30 minutes and will include:
- *practical–related questions*
- *structured questions*
- *short-answer questions.*   *90 marks*

Unit 5
This unit is assessed by means of a written examination paper, which lasts 1 hour 30 minutes and will include:
- *objective questions*
- *structured questions*
- *short-answer questions.*
A third of the marks is related to specified pre-released reading.   *90 marks*

Unit 6
Students will complete an individual investigation.
This is a written report of an experimental investigation, which they have devised and carried out and includes synoptic assessment.
This piece of work will be marked by the teacher and moderated by Edexcel. There is no separate content for this unit.   *50 marks*

The AS/A2 Level Biology course

AS and A2

All Biology GCE A level courses currently studied are in two parts: AS and A2, with three separate units in each.

Some of the units are assessed by written papers, externally marked by the Awarding Body. Some units involve internal assessment of practical skills (subject to moderation).

Each Awarding Body has a common core of subject content in AS and A2. Beyond the common core material, the Awarding Bodies have included more varied content. This study guide contains the common core material and the additional material that is relevant to the Edexcel A2 specification.

In using this study guide, some students may have already completed the AS part of the course. Knowledge of AS is assumed in the A2 part of the course. It is therefore important to revisit the AS information when preparing for the A2 examinations.

What are the differences between AS and A2?

There are three main differences:

(i)   A2 includes the more **demanding** concepts. (Understanding will be easier if you have completed the AS Biology course as a 'stepping stone'.)

(ii) There is a much greater emphasis on the skills of **application** and **analysis** than in AS. (Using knowledge and understanding acquired from AS is essential.)

(iii) A2 includes a substantial amount of **synoptic** material. (This is the drawing together of knowledge and skills across the modules of AS and A2. Synoptic investigative tasks and questions involving concepts across the specification are included.)

How will you be tested?

Assessment units

A2 Biology comprises three units. The first two units are assessed by examinations.

The third component involves centre assessed practical assessment.

Centre-based coursework involves a written report of an experimental investigation which is marked by your teacher. The marks can be adjusted by moderators appointed by the awarding body.

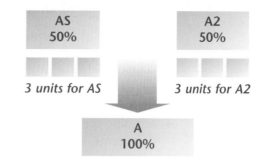

Tests are taken at two specific times of the year, January/February and June. If you are disappointed with a module result, you can resit each unit any number of times. It can be an advantage to you to take a unit test at the earlier optional time because you can re-sit the test. The best mark from each unit will be credited and the lower marks ignored.

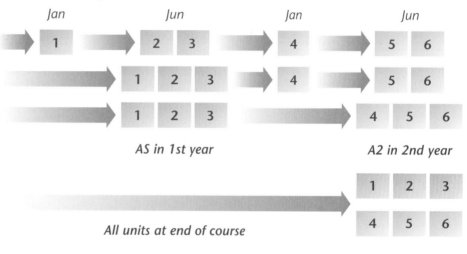

A2 and synoptic assessment

Most students who study A2 have already studied to AS Level. There are three further units to be studied.

Every A Level specification includes synoptic assessment at the end of A2. Synoptic questions draw on the ideas and concepts of earlier units, bringing them together in holistic contexts. Examiners will test your ability to inter-relate topics through the complete course from AS to A2. (See the synoptic chapter page 91).

What skills will I need?

For A2 Biology, you will be tested by assessment objectives: these are the skills and abilities that you should have acquired by studying the course. The assessment objectives shown below.

Knowledge with understanding

- recall of facts, terminology and relationships
- understanding of principles and concepts
- drawing on existing knowledge to show understanding of the responsible use of biological applications in society
- selecting, organising and presenting information clearly and logically

Application of knowledge and understanding, and evaluation

- explaining and interpreting principles and concepts
- interpreting and translating, from one to another, data presented as continuous prose or in tables, diagrams and graphs
- carrying out relevant calculations
- applying knowledge and understanding to familiar and unfamiliar situations
- assessing the validity of biological information, experiments, inferences and statements

You must also present arguments and ideas clearly and logically, using specialist vocabulary where appropriate. Remember to balance your argument!

Investigative and practical skills

You will be expected to carry out an extended practical project. You should plan and devise the experiment yourself but you can discuss your ideas with your teacher. After the investigation, you must produce a written report of between 2700 and 3300 words. This must be word-processed and submitted electronically. It must be entirely your work and include presentation and analysis of your numerical data. The report is marked by your teacher and moderated by Edexcel.

Different types of questions in A2 examinations

Questions in AS and A2 Biology are designed to assess a number of assessment objectives. For the written papers in AS Biology the main objectives being assessed are:

- recall of facts, terminology and inter-relationships
- understanding of principles and concepts and their social and technological applications and implications
- explanation and interpretation of principles and concepts
- interpreting information given as diagrams, photomicrographs, electron micrographs tables, data, graphs and passages
- application of knowledge and understanding to familiar and unfamiliar situations.

In order to assess these abilities and skills a number of different types of question are used.

In A2 Level Biology unit tests these include short-answer questions and structured questions requiring both short answers and more extended answers, together with free-response and open-ended questions.

Short-answer questions

A short-answer question will normally begin with a brief amount of stimulus material. This may be in the form of a diagram, data or graph. A short-answer question may begin by testing recall. Usually this is followed up by questions which test understanding. Often you will be required to analyse data. Short-answer questions normally have a space for your responses on the printed paper. The number of lines is a guide as to the number of words you will need to answer the question. The number of marks indicated on the right side of the papers shows the number of marks you can score for each question part. Here are some examples. (The answers are shown in blue).

The diagram below shows a gastric pit.

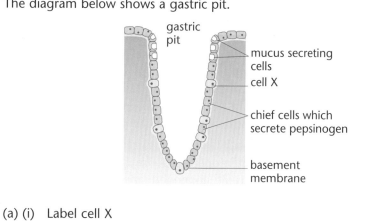

(a) (i) Label cell X   (1)

oxyntic cell

(ii) What is secreted by cell X?   (1)

hydrochloric acid

(b) (i) Protein enters the stomach. What must take place before the hydrolysis of the protein begins?   (2)

Hydrochloric acid acts on pepsinogen, to produce pepsin

(ii) After the protein has been hydrolysed, what is produced?   (1)

polypeptides

Structured questions

Structured questions are in several parts. The parts are usually about a common context and they often progress in difficulty as you work through each of the parts. They may start with simple recall, then test understanding of a familiar or unfamiliar situation. If the context seems unfamiliar the material will still be centred around concepts and skills from the Biology specification. (If a student can answer questions about unfamiliar situations then they display understanding rather than simple recall.)

The most difficult part of a structured question is usually at the end. Ascending in difficulty, a question allows a candidate to build in confidence. Right at the end technological and social applications of biological principles give a more demanding challenge. Most of the questions in this book are structured questions. This is the main type of question used in the assessment of both AS and A2 Biology.

The questions set at A2 Level are generally more difficult than those experienced at AS Level. A2 includes a number of higher-level concepts, so can be expected to be more difficult. The key advice given by this author is:

- Give your answers in greater detail:

    Example: Why does blood glucose rise after a period without food?

    Answer: The hormone glucagon is produced X not enough for credit!)

    The hormone glucagon is produced which results in glycogen breakdown to glucose.

- Look out for questions with a 'sting in the tail'. A2 questions structured questions are less straightforward, so look for a 'twist'. This is identified in the example below.

When answering structured questions, do not feel that you have to complete one question before starting the next. Answering a part that you are sure of will build your confidence. If you run out of ideas go on to the next question. This will be more profitable than staying with a very difficult question which slows down progress. You can return at the end when you have more time.

Extended answers

In A2 and AS Biology questions requiring more extended answers will usually form part of structured questions. They will normally appear at the end of a structured question and will typically have a value of 4 to 20 marks. Longer questions are allocated more lines, so you can use this as a guide as to how many points you need to make in your response. Often for an answer worth 10 marks the mark scheme would have around 12 to 14 creditable answers. You are awarded up to the maximum, 10 marks, in this instance.

Depending on the awarding body, longer, extended questions may be set. These are often open-response questions. These questions are worth up to 20 marks for full credit. Extended answers are used to allocate marks for the **quality of communication**.

Candidates are assessed on their ability to use a suitable style of writing, and organise relevant material, both logically and clearly. The use of specialist biological terms in context is also assessed. Spelling, punctuation and grammar are also taken into consideration. Here is a longer-response question.

Question

Urea, glucose and water molecules enter the kidney via the renal artery. Explain what *can* happen to each of these substances.

In this question one mark is available for communication.   (Total 13 marks)

Urea, glucose and water molecules can pass through the blood capillaries in a glomerulus. ✓ This is as a result of ultrafiltration, ✓ as the narrow diameter of the efferent blood vessel cause a pressure build up. ✓

The three substances pass down the proximal tubule. 100% glucose is reabsorbed in the proximal tubule ✓ so is returned to the blood. Carrier proteins on the microvilli aided by mitochondria, actively transport the glucose across the cells. ✓ Around 80% of the water is reabsorbed in the proximal tubule. ✓ Remaining water and urea molecules continue through the loop of Henlé. Urea continues through the distal tubule to the ureter then the bladder. ✓

More water can be reabsorbed with the help of the countercurrent multiplier. ✓ The ascending limb of the loop of Henlé ✓ actively transports Na^+ and Cl^- ions into the medulla. ✓ Water molecules leave the collecting duct by osmosis due to the ions in the medulla. ✓ Cells of the collecting duct are made more permeable to water by the hormone, ADH. ✓ Some water molecules pass into the capillary network and having been successfully reabsorbed. ✓ Some water molecules continue down the ureters and into the bladder. ✓

Communication mark ✓

Remember that mark schemes for extended questions often exceed the question total, but you can only be awarded credit up to the maximum. In response to this question the candidate would be awarded the maximum of 13 marks which included one communication mark. The candidate gave two more creditable responses which were on the mark scheme, but had already scored a maximum. Try to give more detail in your answers to longer questions. This is the key to A2 success.

Stretch and Challenge

Stretch and Challenge is a concept that is applied to the structured questions in Unit 4 and 5 of the exam papers in A2. In principle, it means that sub-questions become progressively harder so as to challenge more able students and help differentiate between A and A* students.

Stretch and Challenge questions are designed to test a variety of different skills and your understanding of the material. They are likely to test your ability to make appropriate connections between different areas and apply your knowledge in unfamiliar contexts (as opposed to basic recall).

Exam technique

A2 builds from the skills and concepts acquired during the AS course. This Study Guide has been written in a similar style to the AS Biology Guide and incorporates many concepts. The Guide will help you cope as the A2 concepts ascend in difficulty. The chapters explain the ideas in small steps so that understanding takes place gradually. The final aim, of complete understanding of major topics, is more likely.

Can I use my AS Biology Study Guide for A2?

Yes! This will be particularly useful in answering synoptic questions that require direct knowledge of the AS topics.

What are examiners looking for?

Whatever type of question you are answering, it is important to respond in a suitable way. Examiners use instructions to help you to decide the length and depth of your answer. The most common words used are given below, together with a brief description of what each word is asking for.

Define

This requires a formal statement. Some definitions are easy to recall.

Define the term transport.

This is the movement of molecules from where they are in lower concentration to where they are in higher concentration. The process requires energy.

Other definitions are more complex. Where you have problems it is helpful to give an example.

Define the term endemic.

This means that a disease is found regularly in a group of people, district or country.

Use of an example clarifies the meaning. Indicating that malaria is invariably found everywhere in a country confirms understanding.

Explain

This requires a reason. The amount of detail needed is shown by the number of marks allocated.

Explain the difference between resolution and magnification.

Resolution is the ability to be able to distinguish between two points whereas magnification is the number of times an image is bigger than an object itself.

State

This requires a brief answer without any reason.

State one role of blood plasma in a mammal.

Transport of hormones to their target organs.

List

This requires a sequence of points with no explanation.

List the abiotic factors which can affect the rate of photosynthesis in pondweed.

carbon dioxide concentration; amount of light; temperature; pH of water

Describe

This requires a piece of prose which gives key points. Diagrams should be used where possible.

Describe the nervous control of heart rate.

The medulla oblongata ✓ of the brain connects to the sino-atrial node in the right atrium, wall ✓ via the vagus nerve and the sympathetic nerve ✓ the sympathetic nerve speeds up the rate ✓ the vagus nerve slows it down. ✓

Discuss

This requires points both for and against, together with a criticism of each point. (**Compare** is a similar command word).

Discuss the advantages and disadvantages of using systemic insecticides in agriculture.

Advantages are that the insecticides kill the pests which reduce yield ✓ they enter the sap of the plants so insects which consume sap die ✓ the insecticide lasts longer than a contact insecticide, 2 weeks is not uncommon ✓

Disadvantages are that insecticide may remain in the product and harm a consumer e.g. humans ✓ it may destroy organisms other than the target ✓ no insecticide is 100% effective and develops resistant pests. ✓

Suggest

This means that there is no single correct answer. Often you are given an unfamiliar situation to analyse. The examiners hope for logical deductions from the data given and that, usually, you apply your knowledge of biological concepts and principles.

The graph shows that the population of lynx decreased in 1980. Suggest reasons for this.

Weather conditions prevented plant growth ✓ so the snowshoe hares could not get enough food and their population remained low ✓ so the lynx did not have enough hares (prey) to predate upon. ✓ The lynx could have had a disease which reduced numbers. ✓

Calculate

This requires that you work out a numerical answer. Remember to give the units and to show your working, marks are usually available for a partially correct answer. If you work everything out in stages write down the sequence. Otherwise, if you merely give the answer and it is wrong, then the working marks are not available to you.

Calculate the Rf value of spot X. (X is 25 mm from start and solvent front is 100 mm)

$$Rf = \frac{\text{distance moved by spot}}{\text{distance moved by the solvent front}}$$

$$= \frac{25 \text{ mm}}{100 \text{ mm}} = 0.25$$

Outline

This requires that you give only the main points. The marks allocated will guide you on the number of points which you need to make.

Outline the use of restriction endonuclease in genetic engineering.

The enzyme is used to cut the DNA of the donor cell. ✓

It cuts the DNA up like this A T G C C G A T = A T   + G C C G A T ✓
                             T A C G G C T A   T A C G G C   T A

The DNA in a bacterial plasmid is cut with the same restriction endonuclease. ✓

The donor DNA will fit onto the sticky ends of the broken plasmid. ✓

If a question does not seem to make sense, you may have misread it. Read it again!

Some dos and don'ts

Dos

Do *answer the question*

No credit can be given for good Biology that is irrelevant to the question.

Do *use the mark allocation to guide how much you write*

Two marks are awarded for two valid points – writing more will rarely gain more credit and could mean wasted time or even contradicting earlier valid points.

Do *use diagrams, equations and tables in your responses*

Even in 'essay-style' questions, these offer an excellent way of communicating Biology.

Do *write legibly*

An examiner cannot give marks if the answer cannot be read.

Do *write using correct spelling and grammar. Structure longer essays carefully*

Marks are now awarded for the quality of your language in exams.

Don'ts

Don't *fill up any blank space on a paper*

In structured questions, the number of dotted lines should guide the length of your answer.

If you write too much, you waste time and may not finish the exam paper. You also risk contradicting yourself.

Don't *write out the question again*

This wastes time. The marks are for the answer!

Don't *contradict yourself*

The examiner cannot be expected to choose which answer is intended. You could lose a hard-earned mark.

Don't *spend too much time on a part that you find difficult*

You may not have enough time to complete the exam. You can always return to a difficult calculation if you have time at the end of the exam.

What grade do you want?

Everyone would like to improve their grades but you will only manage this with a lot of hard work and determination. You should have a fair idea of your natural ability and likely grade in Biology and the hints below offer advice on improving that grade.

For a Grade A

You will need to be a very good all-rounder.

- You must go into every exam knowing the work extremely well.
- You must be able to apply your knowledge to new, unfamiliar situations.
- You need to have practised many, many exam questions so that you are ready for the type of question that will appear.

The exams test all areas of the syllabus and any weaknesses in your Biology will be found out. There must be no holes in your knowledge and understanding. For a Grade A, you must be competent in all areas.

For a Grade C

You must have a reasonable grasp of Biology but you may have weaknesses in several areas and you will be unsure of some of the reasons for the Biology.

- Many Grade C candidates are just as good at answering questions as the Grade A students but holes and weaknesses often show up in just some topics.
- To improve, you will need to master your weaknesses and you must prepare thoroughly for the exam. You must become a better all-rounder.

For a Grade E

You cannot afford to miss the easy marks. Even if you find Biology difficult to understand and would be happy with a Grade E, there are plenty of questions in which you can gain marks.

- You must memorise all definitions.
- You must practise exam questions to give yourself confidence that you do know some Biology. In exams, answer the parts of questions that you know first. You must not waste time on the difficult parts. You can always go back to these later.
- The areas of Biology that you find most difficult are going to be hard to score on in exams. Even in the difficult questions, there are still marks to be gained. Show your working in calculations because credit is given for a sound method. You can always gain some marks if you get part of the way towards the solution.

What marks do you need?

The table below shows how your average mark is transferred into a grade.

average	80%	70%	60%	50%	40%
grade	A	B	C	D	E

The A* grade

To achieve an A* grade you need to achieve a...

- grade A overall (80% or more on uniform mark scale) for the whole A level qualification
- grade A* (90% or more on uniform mark scale) across your A2 units.

A* grades are awarded for the A level qualification only and not for the AS qualification of individual units.

Four steps to successful revision

Step 1: Understand

- Study the topic to be learned slowly. Make sure you understand the logic or important concepts.
- Mark up the text if necessary – underline, highlight and make notes.
- Re-read each paragraph slowly.

GO TO STEP 2

Step 2: Summarise

- Now make your own revision note summary:
  What is the main idea, theme or concept to be learned?
  What are the main points? How does the logic develop?
  Ask questions: Why? How? What next?
- Use bullet points, mind maps, patterned notes.
- Link ideas with mnemonics, mind maps, crazy stories.
- Note the title and date of the revision notes
  (e.g. Biology: Homeostasis, 3rd March).
- Organise your notes carefully and keep them in a file.

This is now in *short-term memory*. You will forget 80% of it if you do not go to Step 3.
GO TO STEP 3, but first take a 10 minute break.

Step 3: Memorise

- Take 25 minute learning 'bites' with 5 minute breaks.
- After each 5 minute break test yourself:
  Cover the original revision note summary.
  Write down the main points.
  Speak out loud (record on tape).
  Tell someone else.
  Repeat many times.

The material is well on its way to *long-term memory*.
You will forget 40% if you do not do step 4. *GO TO STEP 4*

Step 4: Track/Review

- Create a Revision Diary (one A4 page per day).
- Make a revision plan for the topic, e.g. 1 day later, 1 week later, 1 month later.
- Record your revision in your Revision Diary, e.g.
  Biology: Homeostasis, 3rd March 25 minutes
  Biology: Homeostasis, 5th March 15 minutes
  Biology: Homeostasis, 3rd April 15 minutes
  ... and then at monthly intervals.

Chapter 1

Energy for life

The following topics are covered in this chapter:

- *Metabolism and ATP*
- *Respiration*

- *Autotrophic nutrition*

1.1 Metabolism and ATP

After studying this section you should be able to:

- *understand the principles of metabolic pathways*
- *understand the importance of ATP*

LEARNING SUMMARY

Metabolic pathways

Edexcel   4.5.5, 5.7.7

Inside a living organism there are many chemical reactions occurring at the same time. They may be occurring in the same place, in different parts of the cell or in different cells. Each reaction is controlled by a different enzyme.

> **All the chemical reactions occurring in an organism are called metabolism.**
>
> KEY POINT

Often a number of chemical reactions are linked together. The product of one reaction acts as the substrate for the next reaction. This is called a **metabolic pathway** and each of the chemicals in the pathway are called **intermediates**.

A is the substrate for this pathway, B, C and D are intermediates and E is the product. The enzymes a, b, c and d each control a different step.

$$A \xrightarrow{a} B \xrightarrow{b} C \xrightarrow{c} D \xrightarrow{d} E$$

Metabolic reactions can be classed as one of two types. Reactions that break down complex molecules are called **catabolic reactions** or catabolism. Other reactions build up complex molecules from simple molecules. They are **anabolic reactions** or anabolism.

Anabolic reactions tend to require energy, whereas catabolic reactions release energy. The link between these two types of reactions is a molecule called **ATP**.

Adenosine triphosphate (ATP)

The breakdown of many organic molecules can release large amounts of energy. Similarly, making complex molecules such as proteins requires energy. These reactions must be coupled together. This is achieved by using **adenosine triphosphate (ATP)** molecules.

ATP is a **phosphorylated nucleotide**. (Recall the structure of DNA which consists of nucleotides.) Each nucleotide consists of an organic base, ribose sugar and phosphate group. ATP is a nucleotide with two extra phosphate groups. This is the reason for the term 'phosphorylated nucleotide'.

adenine ——— ribose ——— phosphate ——— phosphate ——— phosphate

ATP is produced from adenosine diphosphate and a phosphate group. This requires energy. The energy is trapped in the ATP molecule. An enzyme called ATPase catalyses this reaction.

			ATP synthase		
ADP	+	P	→	ATP	
adenosine diphosphate		phosphate		adenosine triphosphate	

The hydrolysis of the terminal phosphate group liberates the energy. This can then be used in a number of different ways.

ATPase is a hydrolysing enzyme so that a water molecule is needed, but this is not normally shown in the equation.

	ATPase						
ATP	→	ADP	+	P	+	energy
adenosine triphosphate		adenosine diphosphate		phosphate		

> **KEY POINT**
>
> ATP is the cell's energy currency. A cell does not store large amounts of ATP but it uses it to transfer small packets of energy from one set of reactions to another.

Uses of ATP

- muscle contraction
- active transport
- synthesis of macromolecules
- stimulating the breakdown of substrates to make even more ATP for other uses

1.2 Autotrophic nutrition

> **LEARNING SUMMARY**
>
> After studying this section you should be able to:
> - describe the part played by chloroplasts in photosynthesis
> - recall and explain the biochemical processes of photosynthesis
> - relate the properties of chlorophyll to the absorption and action spectra

The chloroplast

Edexcel   4.5.2

Chloroplasts are organelles in plant cells which photosynthesise. In a leaf, they are strategically positioned to absorb the maximum amount of light energy. Most are located in the palisade mesophyll of leaves, but they are also found in both spongy mesophyll and guard cells. There is a greater amount of light entering the upper surface of a leaf so the palisade tissues benefit from a greater chloroplast density.

The diagram on the following page shows the structure of a chloroplast.

Remember that not all light reaching a leaf may hit a chloroplast. Photons of light can be reflected or even absorbed by other parts of the cell. Around 4% of light entering an ecosystem is actually utilised in photosynthesis!

Even when light reaches the green leaf, not all energy is fixed in the carbohydrate product. Just one quarter becomes chemical energy in carbohydrate.

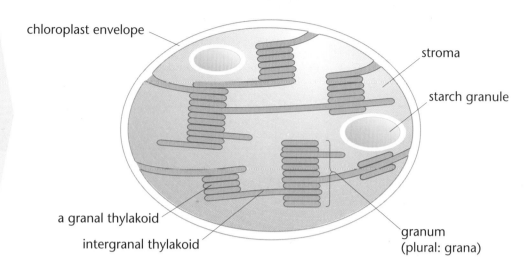

Structure and function

A system of **thylakoid membranes** is located throughout the chloroplast. These are flattened membranous vesicles which are surrounded by a liquid-based matrix, the **stroma**.

Along the thylakoid membranes are key substances:

- chlorophyll molecules
- other pigments
- enzymes
- electron acceptor proteins.

Throughout the chloroplasts, circular thylakoid membranes stack on top of each other to form **grana**. Grana are linked by longer **intergranal thylakoids**. Granal thylakoids and intergranal thylakoids have different pigments and proteins. Each type has a different role in photosynthesis.

The key substances in the thylakoids occur in specific groups comprising pigment, enzyme and electron acceptor proteins. There are two specific groups known as **photosystem I** and **photosystem II**.

Do not be confused by the photosystems. They are groups of chemicals which harness light and pass on energy. Remember this information to understand the biochemistry of photosynthesis.

The photosystems

Each photosystem contains a large number of chlorophyll molecules. As light energy is received at the chlorophyll, electrons from the chlorophyll are boosted to a higher level and energy is passed to pigment molecules known as the **reaction centre**.

> The reaction centre of photosystem I absorbs energy of wavelength 700 nanometres. The reaction centre of photosystem II absorbs energy of wavelength 680–690 nanometres. In this way, light of different wavelengths can be absorbed.
>
> **KEY POINT**

The process of photosynthesis

Edexcel 4.5.3–4/6

The process of photosynthesis is summarised by the flow diagram below.

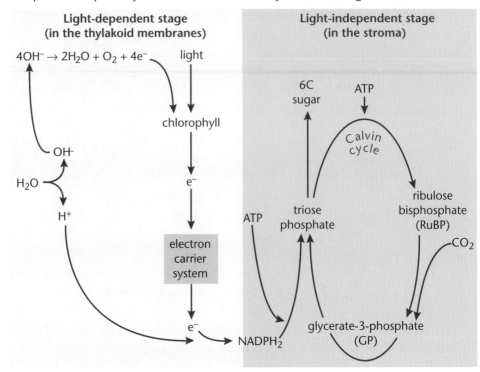

In examinations, look out for parts of this diagram. There may be a few empty boxes where a key substance is missing. Will you be able to recall it?

- Photosynthesis harnesses solar energy.
- Photosynthesis involves light-dependent and light-independent reactions.
- Photosynthesis results in the flow of energy through an ecosystem.

Light-dependent reaction

- Light energy results in the excitation of electrons in the **chlorophyll**.
- These electrons are passed along a series of electron acceptors in the thylakoid membranes, collectively known as the **electron carrier system**.
- Energy from excited electrons funds the production of **ATP** (adenosine triphosphate).
- The final electron acceptor is **NADP⁺**.
- Electron loss from chlorophyll causes the splitting of water (photolysis):

$$H_2O \rightarrow H^+ + OH^- \quad \text{then} \quad 4OH^- \rightarrow 2H_2O + O_2 + 4e^-$$

- Oxygen is produced, water to re-use, and electrons stream back to replace those lost in the chlorophyll.
- Hydrogen ions (H^+) from photolysis, together with $NADP^+$ form **NADPH$_2$**.

No ATP and NADPH$_2$ in a chloroplast would result in no glucose being made. Once supplies of ATP and NADPH$_2$ are exhausted then photosynthesis is ended. In examinations look out for the 'lights out' questions where the light-independent reaction continues for a while until stores of ATP, NADPH$_2$ and GP are used up. These questions are likely to be graph based.

Light-independent reaction

- Two useful substances are produced by the light-dependent stage, ATP and NADPH$_2$. These are needed to drive the light-independent stage.
- They react with glycerate-3-phosphate (GP) to produce a triose sugar – **triose phosphate**.
- Triose phosphate is used *either* to produce a 6C sugar *or* to form **ribulose bisphosphate** (RuBP).
- The conversion of triose phosphate (3C) to RuBP occurs in the Calvin cycle and utilises ATP, which supplies the energy required.
- A RuBP molecule (5C) together with a carbon dioxide molecule (1C) forms two GP molecules (2 × 3C) to complete the Calvin cycle.
- The GP is then available to react with ATP and NADPH$_2$ to synthesise more triose sugar or RuBP.

The enzyme that catalyses the addition of CO_2 to RuBP is called RUBISCO and is often described as the most common protein in the world.

How do the photosystems contribute to photosynthesis?

This can be explained in terms of the **Z scheme** shown below.

The **Z scheme**, so called because the paths of electrons shown in the diagram are in a 'Z' shape.

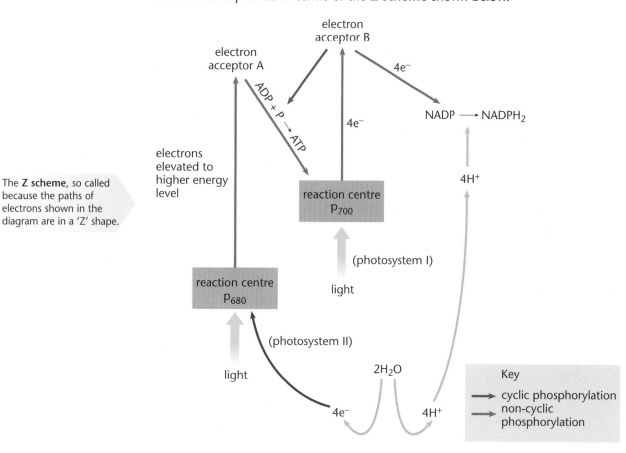

Non-cyclic photophosphorylation

- Light reaches the chlorophyll of both photosystems (P_{680} and P_{700}) which results in the excitation of electrons.
- Electron acceptors receive these electrons (**accepting** electrons is **reduction**).
- P_{680} and P_{700} have become oxidised (**loss** of electrons is **oxidation**).
- P_{680} receives electrons from the **lysis** (splitting) of water molecules and becomes neutral again (referred to as 'hydro lysis').
- Lysis of water molecules releases oxygen which is given off.
- Electrons are elevated to a higher energy level by P_{680} to electron acceptor A and are passed along a series of electron carriers to P_{700}.
- Passage along the electron carrier system funds the production of ATP.
- The electrons pass along a further chain of electron carriers to NADP, which becomes reduced, and at the same time this combines with H^+ ions to form $NADPH_2$.

After analysing this information you will be aware that in cyclic photophosphorylation P_{700} donates electrons then some are recycled back, hence 'cyclic'. In non-cyclic photophosphorylation P_{680} electrons ultimately reach NADP never to return! Neutrality of the chlorophyll of P_{680} is achieved utilising electrons donated from the splitting of water. Different electron sources hence non-cyclic.

Cyclic photophosphorylation

- Electrons from acceptor B move along an electron carrier chain to P_{700}.
- Electron passage along the electron carrier system funds the production of ATP.

1 In a chloroplast, where do the following take place
   (a) light-dependent reaction
   (b) light-independent reaction?

2 (a) Which features do photosystems I and II share in a chloroplast?
   (b) Which photosystem is responsible for:
      (i) the elevation of electrons to their highest level
      (ii) acceptance of electrons from the lysis (splitting) of water?

1 (a) thylakoid membranes (b) stroma.
2 (a) Each photosystem contains a large number of chlorophyll molecules. Light energy is received at the chlorophyll where electrons are boosted to a higher level. Energy is passed to pigment molecules known as the **reaction centre**. The reaction centre of each photosystem absorbs energy (but of different wavelengths).
(b) (i) photosystem I (ii) photosystem II.

1.3 Respiration

After studying this section you should be able to:

- *recall the structure of mitochondria and relate structure to function*
- *understand that respiration liberates energy from organic molecules*
- *explain the stages of glycolysis and Krebs cycle*
- *explain the stages in the hydrogen carrier system*

LEARNING SUMMARY

The site of respiration

Edexcel 5.7.5

Respiration is vital to the activities of every living cell. Like photosynthesis it is a complicated metabolic pathway. The aim of respiration is to break down **respiratory substrates** such as glucose to produce **ATP**.

Respiration consists of a number of different stages. These occur in different parts of the cell. Some stages require oxygen and some do not.

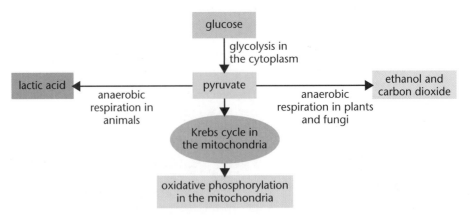

Glycolysis occurs in the cytoplasm of the cell. The pyruvate produced then enters the mitochondria. The Krebs cycle then occurs in the matrix of the mitochondria followed by oxidative phosphorylation which occurs on the inner membrane of the cristae.

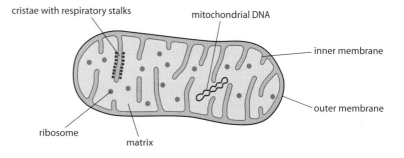

cristae with respiratory stalks

mitochondrial DNA

inner membrane

outer membrane

ribosome

matrix

The biochemistry of respiration

Edexcel   5.7.8–5.7.10

Glycolysis and the Krebs cycle

Both processes produce ATP from substrates but the Krebs cycle produces **many more** ATP molecules than glycolysis. Every stage in each process is catalysed by a specific enzyme. In aerobic respiration, **both** glycolysis and the Krebs cycle are involved, whereas in anaerobic respiration only glycolysis takes place.

The flow diagram below shows stages in the breakdown of glucose in glycolysis and the Krebs cycle. The flow diagram shows only the main stages of each process.

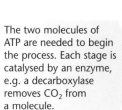

The two molecules of ATP are needed to begin the process. Each stage is catalysed by an enzyme, e.g. a decarboxylase removes CO_2 from a molecule.

The production of hydrogen atoms during the process can be monitored using DCPIP (dichlorophenol indophenol). It is a hydrogen acceptor and becomes colourless when fully reduced.

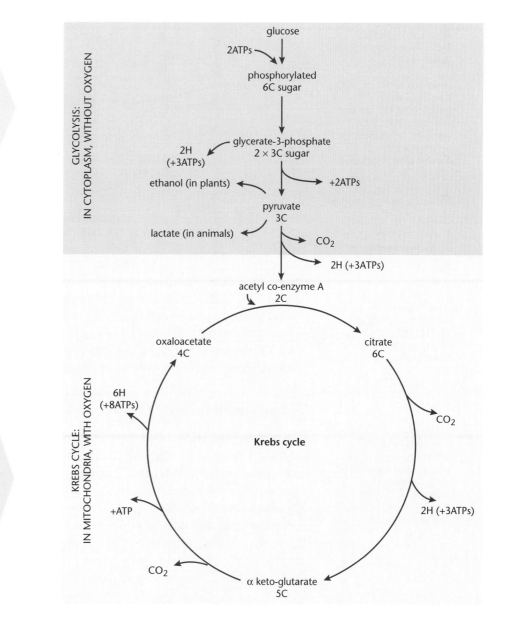

The flow diagram shows that glycolysis produces 2 × 2ATP molecules but uses 2ATP so the net production is 2ATP. The Krebs cycle makes 2ATP directly. All the rest of the ATP molecules that are made (shown in brackets) are produced in oxidative phosphorylation.

Oxidative phosphorylation

The main feature of this process is the electron carrier or electron transport system. The hydrogen that is given off by glycolysis and the Krebs cycle is picked up by acceptor molecules such as **NAD**. These hydrogen atoms are passed along a series of carriers on the inner membrane of the mitochondrion.

Oxidation	Reduction
gain of oxygen	loss of oxygen
loss of hydrogen	gain of hydrogen
loss of electrons	gain of electrons

Electron transport system

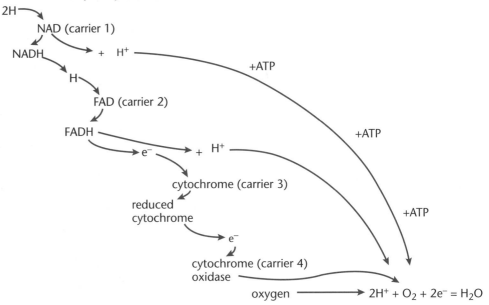

When oxidation takes place then so does reduction, simultaneously, e.g. NADH$_2$ passes H to FAD. The NAD loses hydrogen and as a result becomes oxidised. FAD gains hydrogen and becomes FADH$_2$, and is therefore reduced. The generic term for an enzyme which catalyses this is **oxidoreductase**. Additionally an enzyme which removes hydrogen from a molecule is a **dehydrogenase**. The result is that three ATPs are produced every time 2H atoms are transported.

The chemiosmotic theory

It has now been shown that the carrier molecules are arranged on the membrane of the cristae in a specific way. This means that hydrogen ions are moved out of the matrix and into the space between the two membranes. This sets up a pH gradient. The hydrogen ions can re-enter the matrix through the respiratory stalks. This movement is linked to ATP production and this process is called the **chemiosmotic theory**.

The maximum ATP yield per glucose molecule is:
GLYCOLYSIS 2
KREBS CYCLE 2
OXIDATIVE PHOSPHORYLATION 34
= 38 ATP

Oxygen is needed at the end of the carrier chain as a hydrogen acceptor. This is why we need oxygen to live. Without it, the generation of ATP along this route would be stopped.

This is sometimes known as the hydrogen carrier system.

The carrier, NAD, is nicotinamide adenine dinucleotide. Similarly, FAD is flavine adenine dinucleotide.

Hydrogen is not transferred to cytochrome. Instead, the 2H atoms ionise into 2H$^+$ + 2e$^-$. H is passed via an intermediate co-enzyme Q to cytochrome.

Only the electrons are carried via the cytochromes.

e$^-$ is an electron.
H$^+$ is a hydrogen ion or proton.

An enzyme can be both an oxidoreductase and a dehydrogenase at the same time!

This theory also explains ATP production in photophosphorylation in the chloroplast. The only difference is that the ions are moved in the opposite direction.

Anaerobic respiration

If oxygen is in short supply then the final hydrogen acceptor for the hydrogen atoms is missing. This means that oxidative phosphorylation will stop and NAD will not be regenerated. This will result in the Krebs cycle being unable to function.

Glycolysis can continue and produce 2ATP molecules but it would soon run out of NAD as well. A small amount of NAD can be regenerated by converting the pyruvate to lactate or ethanol. This allows glycolysis to continue in the absence of oxygen. This is anaerobic respiration.

Ethanol is produced in plants and yeast, lactate is made in animals.

> **KEY POINT**
>
> Anaerobic respiration will make 2ATP molecules from one glucose molecule compared to a possible 38ATP in aerobic respiration.

When oxygen becomes available again, the lactate is converted back to pyruvate in the liver.

Progress check

1 Explain how hydrogen atom production in cells during aerobic respiration results in the release of energy for cell activity.

2 Give **three** similarities between respiration and photosynthesis.

3 (a) Name the **four** carriers in the electron transport system in a mitochondrion. Give them in the correct sequence.

   (b) Name the waste product which results from the final stage of the electron transport system.

4 For each of the following statements indicate whether a molecule would be oxidised or reduced.

   (a) (i)   loss of oxygen
       (ii) gain of hydrogen
       (iii) loss of electrons

   (b) Which type of enzyme enables hydrogen to be transferred from one molecule to another?

1 Used in the electron transport system to produce ATP; 3ATP molecules produced for every 2H atoms produced; ATP → ADP + P + energy released

2 The stages of each process are catalysed by enzymes; both processes involve ATP; respiration involves GP in glycolysis and photosynthesis involves GP in the light-independent stage

3 (a) NAD → FAD → cytochrome → cytochrome oxidase
   (b) water

4 (a) (i) reduced (ii) reduced (iii) oxidised
   (b) Oxidoreductase/dehydrogenase

Sample question and model answer

Radioactivity is used to label molecules. They can then be tracked with a Geiger-Müller counter.

In an experiment, pondweed was immersed in water which was saturated with radioactive carbon dioxide ($^{14}CO_2$). It was illuminated for a time so that photosynthesis took place, the light was then switched off. The graph below shows the relative levels of some substances.

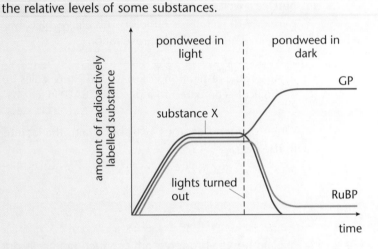

Always be ready to link the rise in one graph line with the dip of another. The relationship holds true here as substance X and RuBP are used up in the production of GP via the Calvin cycle. It is likely that some GP would have been used with substance X to make triose sugar. This is not shown on this graph.

Use the graph and your knowledge to answer the following questions.

(a) (i) Substance X is produced after a substance becomes reduced during the light-dependent stage of photosynthesis. Name substance X.   [1]

NADPH$_2$, reduced nicotinamide adenine dinucleotide phosphate

(ii) Explain why substance X cannot be produced without light energy.   [3]

· Light energy removes electrons from chlorophyll.
· The electrons are passed along the electron carrier chain.
· The electrons are needed to reduce NADP.

(b) Explain the levels of substance X, GP and RuBP after the lights were turned off.   [6]

· It seems that substance X is used to make the other two substances because it becomes used up.
· Supply of substance X cannot be produced without light energy.
· GP is made from RuBP.
· GP levels out because more NADPH$_2$ is needed to make triose sugar or RuBP, the supply being exhausted.
· RuBP levels out at a low level because more NADPH$_2$ is needed to make GP.
· ATP is needed to make RuBP, ATP is needed to make GP.

ATP is not shown on the graph. Always be ready to consider substances involved in a process but not shown. Here it is worth a mark to remember that ATP is needed to continue the light-independent system of photosynthesis.

(c) After the lights were switched off glucose was found to decrease rapidly.
Explain this decrease.   [1]

· Glucose is used up in respiration to release energy for the cell.

(d) Give the specific sites of each of the following stages of photosynthesis in a chloroplast: [2]

   (i) light-dependent stage   thylakoid membranes
   (ii) light-independent stage.   stroma

Practice examination questions

1 The flow diagram below shows stages in the process of glycolysis.

            2ATPs

glucose → phosphorylated → GP → substance X → lactate
6C sugar   6C sugar   glycerate-   3C
                                3-phosphate
                               (2 × 3C)
                               2ATPs

Use the information in the diagram and your knowledge to answer the questions below.

(a) Where in a cell does the above process take place?   [1]

(b) Name substance X.   [1]

(c) How many ATPs are *produced* during the above process?   [1]

(d) Is the above process from an animal or plant?
    Give a reason for your answer.   [1]

(e) Under which condition could lactate be metabolised?   [1]

[Total: 5]

2 The flow diagram below shows part of the electron carrier system in an animal cell.

FADH → FAD + H^+ + e^-

(a) Where in a cell does this process take place?   [1]

(b) From which molecule did FAD receive H to become FADH?   [1]

(c) Which molecule receives the electron produced by the breakdown of FADH?   [1]

(d) As FADH becomes oxidised a useful substance is produced.
    Name the substance.   [1]

[Total: 4]

Chapter 2

Response to stimuli

The following topics are covered in this chapter:

- Neurone structure and function
- Coordination by the CNS
- Plant sensitivity
- Receptors
- Response

2.1 Neurone structure and function

After studying this section you should be able to:

LEARNING SUMMARY

- describe the structure of a motor neurone, a sensory neurone and a relay neurone.
- understand the function of sensory, motor and relay neurones
- understand nervous transmission by action potential
- describe the mechanisms of synaptic transmission

The structure and function of neurones

Edexcel ▶ 5.8.3

Neurones are **nerve cells** which help to coordinate the activity of an organism by transmitting **electrical impulses**. Many neurones are usually gathered together, enclosed in connective tissue to form **nerves**.

> Important features of neurones.
>
> 1 Each has a **cell body** which contains a nucleus.
> 2 Each communicates via processes **from the cell body**.
> 3 Processes that carry impulses away from the cell body are known as **axons**.
> 4 Processes that carry impulses towards the cell body are known as **dendrons**.
> 5 All neurones transmit **electrical impulses**.

KEY POINT

The nervous system consists of a range of different neurones which work in a network through the organs. The diagrams show three types of neurone.

Notice the direction of the impulse and that motor neurones have long axons and short dendrons. This is the other way round for sensory neurones.

Key points from AS

- **The cell surface membrane**
  Revise AS pages 40–41
- **The movement of molecules in and out of cells**
  Revise AS pages 42–43
- **The specialisation of cells**
  Revise AS pages 31–32

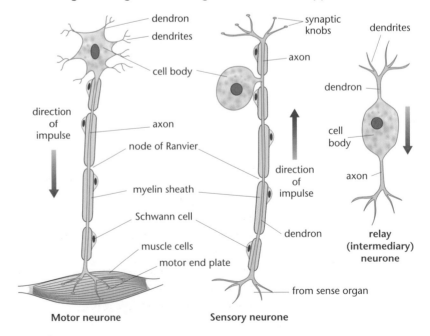

Motor neurone Sensory neurone

Myelinated neurones

Sensory and motor neurones are examples of myelinated neurones. This enables them to transmit an impulse at a greater velocity. Myelinated neurones have the following characteristics:

- The axon or dendron is insulated by a **myelin sheath**.
- The myelin sheath is formed by a **Schwann cell** wrapping around the axon many times. This forms many layers of cell membrane surrounding the axon.
- At intervals there are gaps in the sheath, between each Schwann cell, called **nodes of Ranvier**.

Cross-section of axon

The myelin sheath is often called a 'fatty' sheath because it is made of many layers of cell membrane which are composed largely of phospholipids.

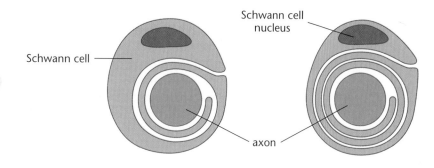

What are the roles of the sensory and motor neurones?

There are many similarities between the structure of sensory and motor neurones but they have different functions:

- The sensory neurones transmit impulses towards the central nervous system (CNS) from the receptors.
- The motor neurones transmit impulses from the CNS to effectors, such as muscles, to bring about a response.
- Relay neurones may form connections between sensory and motor neurones in the CNS.

Transmission of an action potential along a neurone

Edexcel 5.8.4

Neurones can 'transmit an electrical message' along an axon. However, you must never write this in your answers. Instead of nerve impulse, you must now use the term **action potential**.

The diagrams below show the sequence of events which take place along an axon as an action potential passes.

Resting potential

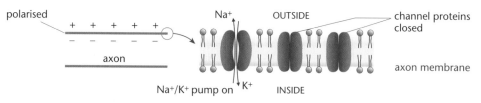

- There are 30 times more Na$^+$ ions on the outside of an axon during a resting potential.
- If any Na$^+$ ions diffuse in, then they are expelled by the '**sodium–potassium pump**'.
- The 'sodium–potassium pump' is an active transport mechanism by which a carrier protein, with ATP, expels Na$^+$ ions against a concentration gradient and allows K$^+$ ions into the axon.

Under resting conditions, the membrane of the axon is fairly impermeable to sodium ions.

- This creates a **polarisation**, i.e. there is a +ve charge on the outside of the membrane and a –ve charge on the inside.
- The potential difference is called the **resting potential** and can be measured at around –70 millivolts.

Action potential – depolarisation

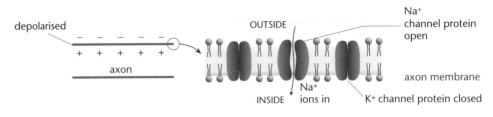

- During an action potential sodium channel proteins open to allow Na+ ions into the axon.
- There is now a –ve charge on the outside and a +ve charge on the inside known as **depolarisation**.
- The potential difference changes to around +50 millivolts.
- The profile of the action potential, shown by an oscilloscope, is always the same.

Action potential – repolarisation

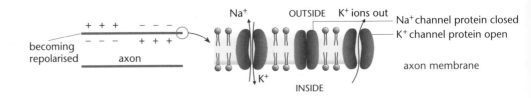

- A K^+ channel opens so K^+ ions leave the axon.
- This results in the membrane becoming polarised again.
- Any Na^+ ions that have entered during the action potential will be removed by the 'sodium–potassium pump'.

Measuring an action potential

- The speed and profile of action potentials can be measured with the help of an oscilloscope.
- The profile of the action potential for an organism always shows the same pattern, like the one shown.
- The changes in potential difference are tracked via a time base.
- Using the time base you can work out the speed at which action potentials pass along an axon as well as how long one lasts.

The diagram shows the typical profile of an action potential.

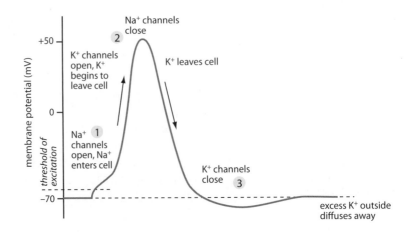

- The front of the action potential is marked by the Na⁺ channels in the membrane opening.
- The potential difference increases to around +50 millivolts as the Na⁺ ions stream into the axon.
- The Na⁺ channels then close and K⁺ channels in the membrane open.
- K⁺ ions leave the axon and the membrane repolarises.
- During the **refractory period** no other action potential can pass along the axon, which makes each action potential separate or discrete.

Saltatory conduction

The reason why myelinated neurones are faster than non-myelinated neurones is that the action potential 'jumps' from one node of Ranvier to the next. This is because this is the only place where Na⁺ ions can pass across the membrane. This is called **saltatory conduction**.

Progress check

What is the function of each of the following?

(a) receptor
(b) axon
(c) myelin sheath
(d) terminal dendrites

(d) Terminal dendrites have motor end plates which can stimulate muscle tissue to contract.
(c) Myelin sheath is a membrane enclosing fat which acts as an insulator.
(b) Transmit action potential with the help of mitochondria.
(a) Receptors respond to stimulus by producing an action potential.

How do neurones communicate with each other?

Edexcel 5.8.5

The key to links between neurones are structures known as **synapses**. Terminal dendrites branch out from neurones and terminate in **synaptic knobs**. The diagram below shows a synaptic knob separated from an interlinking neurone by a synapse.

Remember an impulse can 'cross' a synapse by chemical means and the route is in ONE direction only. They cannot go back!

A synapse which conducts using acetylcholine is known as a cholinergic synapse.

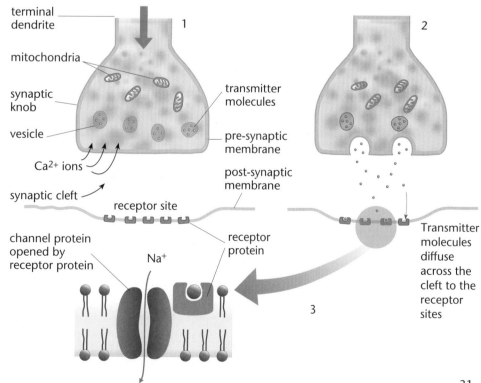

31

There are **two** types of synapse:
* **excitatory** which can stimulate an action potential in a linked neurone
* **inhibitory** which can prevent an action potential being generated.

As an action potential arrives at a synaptic knob, the following sequence takes place

* **Channel proteins** in the **pre-synaptic membrane** open to allow Ca^{2+} ions from the synaptic cleft into the synaptic knob.
* **Vesicles** then merge with the pre-synaptic membrane, so that **transmitter molecules** such as **acetylcholine** are **secreted** into the gap.
* The transmitter molecules diffuse across the cleft and bind with specific **sites** in **receptor proteins** in the **post-synaptic membrane**.
* Every receptor protein then opens a **channel protein** so that ions such as Na^+ pass through the post-synaptic membrane into the cell.
* The Na^+ ions **depolarise** the post-synaptic membrane.
* If enough Na^+ ions enter then depolarisation reaches a **threshold level** and an **action potential** is generated in the cell.
* Enzymes in the cleft then remove the transmitter substance from the binding sites, e.g. **acetylcholine esterase** removes **acetylcholine** by hydrolysing it into choline and ethanoic acid.
* Breakdown products of transmitter substances are absorbed into the synaptic knob for re-synthesis of transmitter.

Remember that the generation of an action potential is ALL OR NOTHING. Either enough Na^+ ions pass through the post-synaptic membrane and an action potential is generated OR not enough reach the other side, and there is no effect.

> ## Summation
>
> A single action potential may arrive at a synaptic knob and result in some transmitter molecules being secreted into a cleft. However, there may not be enough to cause an action potential to be generated. If a series of action potentials arrive at the synaptic knob then the build up of transmitter substances may reach the threshold and the neurone will now send an action potential. We say that the neurone has 'fired' as the action potential is produced.

KEY POINT

Progress check

1 Explain the importance of summation at a synapse.

2 The diagram shows a synaptic knob.

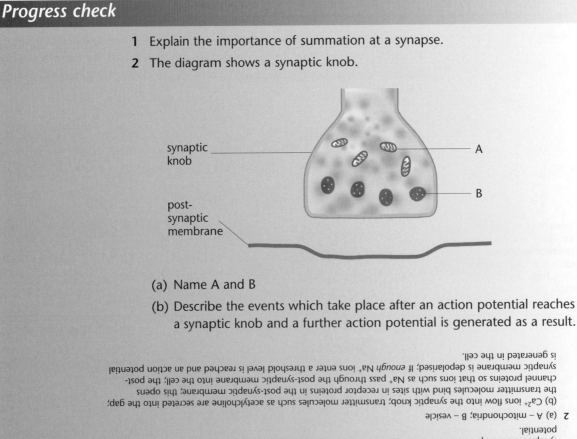

synaptic knob — A

post-synaptic membrane — B

(a) Name A and B

(b) Describe the events which take place after an action potential reaches a synaptic knob and a further action potential is generated as a result.

1 A single action potential may arrive at a synaptic knob; there may not be enough transmitter molecules being secreted into a cleft to cause an action potential to be generated; a series of action potentials arrive at the synapse to build up transmitter substances to reach the threshold; the neurone will now send an action potential.

2 (a) A – mitochondria, B – vesicle
(b) Ca^{2+} ions flow into the synaptic knob; transmitter molecules such as acetylcholine are secreted into the gap; the transmitter molecules bind with sites in receptor proteins in the post-synaptic membrane; this opens channel proteins so that ions such as Na^+ pass through the post-synaptic membrane into the cell; the post-synaptic membrane is depolarised; *if enough* Na^+ ions enter a threshold level is reached and an action potential is generated in the cell.

2.2 Receptors

After studying this section you should be able to:
- *list the different types of receptors*
- *explain how receptors trigger nerve impulses*
- *describe how rods and cones respond to light*

Types of receptors

Edexcel 5.8.6

The function of receptors is to convert the energy from different stimuli into nerve impulses in sensory neurones.

There are a range of different types of sensory cell around the body. Each type responds to different stimuli. Receptors are classified according to these different stimuli:

- **Photoreceptors**, respond to light, e.g. rods and cones in the retina.
- **Chemoreceptors**, respond to chemicals, e.g. taste buds on the tongue.
- **Thermoreceptors**, respond to temperature, e.g. skin thermoreceptors.
- **Mechanoreceptors**, respond to physical deformation, e.g. Pacinian corpuscles in the skin or hair cells in the ear.
- **Proprioreceptors**, respond to change in position in some organs, e.g. in muscles.

The more receptors there are in a position, the more sensitive it is, e.g. the fingers have many more touch receptors than the upper arm.

Did you know?
The umbilical cord has no receptors. It can be cut without any pain.

Stimulation of a receptor usually causes it to depolarise. This is called a **generator potential**. If this change is beyond a certain magnitude, it will trigger an action potential in a sensory neurone.

Some receptors are found individually in the body such as **Pacinian corpuscles** which detect pressure in the skin. Other receptors are gathered together into sense organs. An example of this is the eye which contains receptors called rods and cones.

How do the cells of the eye respond to light?

Edexcel 5.8.6

The retina in the eye contains two types of cell which are **photosensitive**, the rod cells and the cone cells. They each have different properties.

Rod cells

- They are very sensitive to the **intensity of light**, but are not sensitive to colour.
- They can respond to even dim light.
- They respond by the following reaction:
  rhodopsin → opsin + retinal
- They have low visual acuity in dim conditions.
- Opsin opens **ion channels** in the cell surface membrane which can result in the generation of an action potential.
- Rhodopsin can be re-generated during an absence of light.

wavelength/nm

The colour vision mechanism appears to require three types of cone: RED, GREEN, and BLUE and gives the *trichromatic* theory.

Cone cells

- Cone cells require **high light intensities** to be responsive – high visual acuity.
- They respond by the following reaction:
  iodopsin → photopsin + retinal
- They exist in three different types: red, green and blue, each having a different form of iodopsin:
  – **red** cones are stimulated by wavelengths of red light
  – **green** cones are stimulated by wavelengths of green light

– **blue** cones are stimulated by wavelengths of blue light
– all three cones are stimulated by white light

• Opsin again opens ion channels in the membranes which can lead to the generation of an action potential.

Structure of the retina

Remember that red light reaching a cone sensitive to only blue light would not stimulate the generation of an action potential. Cones are only sensitive to light of a specific range of wavelengths.

Note the direction of light as it reaches the outer retinal surface.

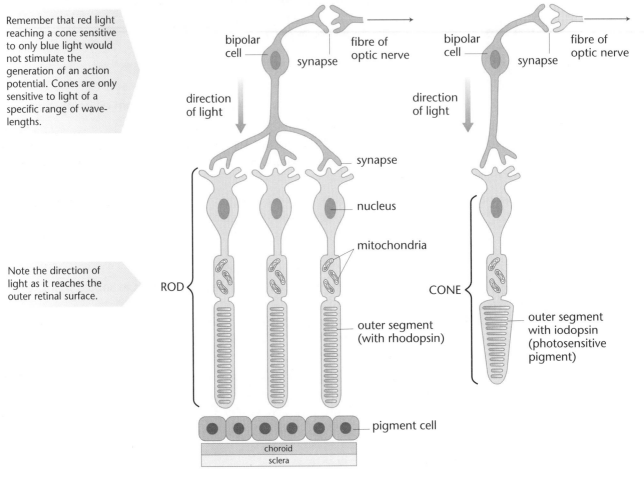

Visual acuity

Did you know that the fovea consists almost entirely of cones?

What do we see with in dim conditions?

You have guessed it, rods. Their ability to detect dim light is useful but there is no colour!

Visual acuity is a measure of the detail we can see. The **cones** are responsible for high visual acuity (**high resolution**). **Large numbers** are packed **very close** to each other in the fovea. **ONE** cone cell synapses onto **ONE** bipolar cell which in turn synapses onto **ONE** ganglion cell as the information is relayed to the visual cortex. Spatially, much more clarity is perceived than for the rods. The image can be likened to a television picture with high numbers of pixels. Compare this with the rods. The rod cells are not packed close together so that visual acuity is low.

Convergence

Many rods can synapse onto one bipolar cell. A ray of light reaching one rod may not be enough to stimulate an action potential along a nerve pathway. **Several** rods link to **one** bipolar cell so that enough transmitter molecules at a synapse reach the threshold level. This depolarisation results in an action potential in the bipolar cell. This is **summation**, as a result of rod cell teamwork!

2.3 Coordination by the CNS

After studying this section you should be able to:

- outline the structure and functions of the brain and spinal cord
- understand the main functions of cerebrum, cerebellum, medulla oblongata and hypothalamus
- describe some of the methods used to study brain function

LEARNING SUMMARY

The structure and functions of the CNS

Edexcel   5.8.9

The CNS consists of the brain and spinal cord which work together to aid the coordination of the organism. The human brain has many functions. The spinal cord takes impulses from the brain to **effectors** and in the opposite direction, impulses from **receptors** are channelled to the brain.

The brain has a complex 3D structure. The diagram below shows part of the brain structure – the major components only.

The CNS is like a motorway with impulses going in both directions.

- **Afferent neurones** take impulses **from** organs **to** the **CNS**.
- **Efferent neurones** take impulses **from** the **CNS** to an **organ**.

Learn these carefully. There are no marks for reversal!

The human brain

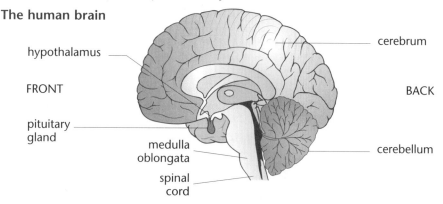

hypothalamus

FRONT

pituitary gland

medulla oblongata

spinal cord

cerebrum

BACK

cerebellum

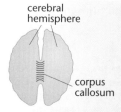

cerebral hemisphere

corpus callosum

There are two cerebral hemispheres: the left and the right. Note that the right hemisphere controls the left side of the body and vice versa.

Alzheimer's disease

Neurones in the cortex of the cerebrum become progressively less able to produce neurotransmitter substances. Acetylcholine and noradrenaline are usually deficient resulting in major personality changes. The cause is often unknown, but can be genetic.

Functions of parts of the brain

Cerebrum

- **Receives sensory information** from many organs, e.g. impulses are sent from the eyes to the visual cortex at the back of the cerebrum.
- **Initiates motor activity** of many organs.
- The front of the cerebrum holds the **memory** and **intelligence** in a network of multi-polar neurones.

Cerebellum

- Has a key role in the coordination of **balance** and smooth, controlled muscular movements.
- Initiation of a movement may be by the cerebrum but the **smooth, well-coordinated** execution of the movement is only possible with the help of the cerebellum.

Medulla oblongata

- Its respiratory centre controls the rhythm of breathing with nerve connections to the intercostal muscles and the diaphragm.
- Its cardiovascular centre controls the cardiac cycle via the sympathetic and vagus nerve.
- Connects to the sino-atrial node of the heart.

The hypothalamus is the key structure in maintaining a homeostatic balance in the body. It is like a thermostat in a house, switching the heating system on or off as internal conditions change. Similarly, it is able to control chemical levels in the blood.

Hypothalamus

- Has an exceptionally rich blood supply.
- Many receptors are located in the blood vessel walls which supply it.
- These receptors are highly sensitive detectors which monitor:
  - temperature
  - carbon dioxide
  - ionic concentration of plasma.
- Controls body temperature by various regulatory mechanisms.
- Controls ADH secretion by the pituitary gland and is, therefore, responsible for the water content of both blood plasma and urine.

Pituitary gland

- Secretes a range of hormones and is the major control agent of the endocrine system.
- Responds to neurosecretion and release factors from the hypothalamus.
- Together with the hypothalamus, it is part of a number of negative feedback loops.
- Is the link between the nervous system and the endocrine system.

The above cover some functions of parts of the brain, but there are many more.

> The human brain consists of approximately 10^{12} neurones and all are present at birth. It is no wonder that a baby's head is proportionally large at this stage of the life cycle. During the first three months after birth, many synaptic connections are made. This is a most important developmental stage. Neurones cannot be replaced once damaged.
>
> **KEY POINT**

Progress check

State **two** functions of each of the following parts of the human brain:

(a) cerebrum (b) cerebellum (c) medulla oblongata.

(a) (i) Receives sensory impulses from the eyes to the visual cortex, enabling sight.
(ii) Controls voluntary motor activity of the leg muscles.
(b) (i) Coordinates balance, e.g. enables upright stance in humans.
(ii) Enables smooth movement, e.g. hitting a golf ball with a club straight down the fairway. (You could swing the club by voluntary control from the cerebrum but smooth coordination is by the cerebellum.)
(c) (i) Controls the rhythm of breathing with nerve connections to the intercostal muscles and the diaphragm.
(ii) Controls the heart rate via the sympathetic and vagus nerve.

Studying brain function

Edexcel   5.8.10–5.8.12

Scientists have used a number of different techniques to try to work out which parts of the brain are responsible for different functions. They are also starting to investigate how the brain might process information.

There are a number of methods that have been used:

Accidents and diseases

Many of the earlier studies involved looking at people or animals where visible damage has occurred to parts of the brain. By observing the effects on the organism, it is possible to draw some simple conclusions about the functions of parts of the brain. An example was Phineas Gage who survived damage to his cerebrum caused by an iron bar. The damage caused a change in his personality.

The human brain

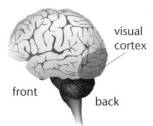

visual cortex

front

back

Hubel and Weisel won the Nobel Prize for this work in 1981.

Animal models

Animals have been used to investigate brain function. This may involve observing the effects of damage or directly measuring the response of neurones in the brain by using electrodes. The second of these ideas was used by the scientists David Hubel and Torsten Weisel in the 1960s using monkeys and cats.

Hubel and Weisel recorded the responses of neurones in the part of the cerebral hemisphere called the visual cortex.

The animal was watching a screen and they found that the neurones depolarised only when a bar was moving across the screen at a certain angle. They realised that the bar had to be moving and that different cells responded when it moved at different angles. This was a major step in trying to discover how the brain processes the information that enters the eyes.

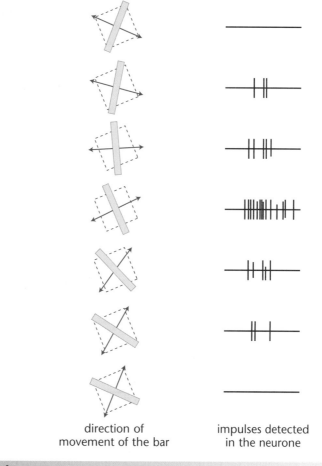

direction of movement of the bar

impulses detected in the neurone

Brain scans

There are now a number of different types of scans that can be performed on living brains. The results can be used to study brain function and are important in medical diagnosis.

Magnetic resonance imaging (MRI) scans use strong magnetic fields and radio waves to produce a detailed image of the brain. The scanner consists of a large tube that contains a series of powerful magnets. The patient lies inside the tube during the scan.

fMRI is a special type of MRI scan that measures the activity in parts of the brain. It can detect which neurones are more active because they use up more oxygen.

Computerised tomography (CT or CAT scan) takes a series of X-ray images and uses a computer to put them together. The CT machine takes pictures of the brain from different angles and gives a series of images of cross sections or 'slices' through the brain.

2.4 Response

After studying this section you should be able to:

- *describe different types of response*
- *understand how neurones function together in a reflex arc*
- *outline the features of the autonomic nervous system*
- *describe the structure of skeletal muscle and understand the sliding filament mechanism*

LEARNING SUMMARY

Different types of response

Edexcel ▷ 5.8.7

In order to bring about a response, nerve impulses are sent to effectors via motor neurones. Some responses do not require conscious thought. These are called reflexes or reflex actions.

The reflex arc

How can we react quickly without even thinking about making a response?

Often the brain is not involved in the response so the time taken to respond to a stimulus is reduced. This rapid, automatic response is made possible by the **reflex arc**.

A reflex arc

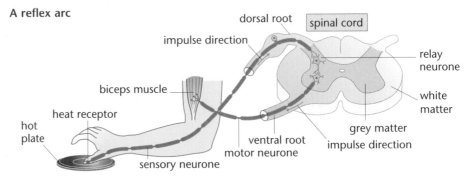

Features of a reflex arc

- The stimulus elicits a response in a **receptor**.
- As a result, an **action potential** is generated along a sensory neurone.
- The **sensory neurone** enters the spinal cord via the **dorsal root** and **synapses** onto a **relay neurone**.
- This intermediate neurone synapses onto a **motor neurone** which in turn conducts the impulse to a muscle via its **motor end plates**.
- The muscle contracts and the arm instantly **withdraws** from the stimulus before any harm is done.
- The complete list of events takes place so quickly because the impulses do not, initially, go to the brain! The complete pathway to the muscle conducts the impulse so rapidly, before the brain receives any sensory information.

> It is other afferent neurones which *finally* take impulses to the brain enabling us to be aware of the arc which has just taken place. These afferent neurones are NOT part of the reflex arc.

> Reflexes have a high survival value because the organism is able to respond so rapidly. Additionally, they are always automatic. There are a range of different reflexes, e.g. iris/pupil reflex and saliva production.

KEY POINT

The iris/pupil reflex

radial muscles relax

circular muscles contract

pupil constricts

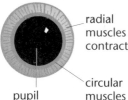

radial muscles contract

circular muscles relax

pupil dilates

This response does involve the brain, but because conscious thought is not involved, it is still classed as a reflex.

The diagrams in the margin show the two extremes of pupil size.

- The amount of light entering the eye is detected by **receptors** in the **retina**.
- Reflex pathways lead to the **circular** and **radial** muscles of the **iris**.
- **High-intensity** light activates the **circular muscles** of the iris to **contract**; as the radial muscles relax so the **pupil gets smaller**. (The advantage of this is too much light does not enter and so does not damage the retina.)
- **Low-intensity** light activates the **radial muscles** of the iris to **contract**; as the circular muscles relax so the **pupil gets wider**. (The advantage of this is that the eye allows enough light to see.)
- A balance between the two extremes is achieved across a gradation of light conditions.

Learned responses

Simple reflexes are innate responses that animals are born with. However, as soon as an animal is born it will start to develop learned responses. These are patterns of behaviour that are modified by past experience. An example of a type of learned response is habituation.

Police horses are trained using habituation so that they ignore loud noises and visual distractions.

This takes place when an organism is subjected to a stimulus which is not harmful or rewarding. As a result of continued subjection to a stimulus, a response will gradually decrease and can finally disappear completely. A farmer puts an electronic bird scarer into a field. Birds are frightened off by frequent 'bangs'. They return, gradually getting closer and finally learn that the scarer is non-threatening. Soon, they feed close to the scarer, which has no effect. This is called **habituation**.

Advertisements have a short 'shelf-life'. Continued exposure to the same advertisement results in habituation so that the response decreases. This is why media advertising is replaced every few weeks.

Nature v nurture

As soon as we are born we start to learn. Our behaviour is therefore modified by our experiences and our environment.

For hundreds of years, the scientists have been arguing about whether it is our genes or the environment which has a greater effect on our development.

> This debate is called *nature versus nurture* or the *nature–nurture debate*.

KEY POINT

Initially, scientists looked at family trees and argued that features such as intelligence were inherited. This, however, ignored the effect of the environment.

Now, more scientific methods are used:

Twin studies: These studies ideally use identical twins that have been separated at birth and brought up in different environments. This means that they have the same genes and so the impact of the environment can be assessed. One famous study on twins is called the Minnesota Twin Study. When scientists looked at identical twins that had been separated they found that the twins' personalities were very similar despite the different upbringing.

Animal studies: Obviously, identical twins that are separated at birth are quite rare. It is also unethical to control their environment. Therefore, many studies have been undertaken on animals. Animals can be taken from the same litter and their environments altered and controlled. For example, in laboratory animals, there is more brain growth in animals kept in environments full of toys and other animals, compared to animals kept alone in a bare cage.

Cross-cultural studies: These studies look at the differences between children who have been brought up in different cultural backgrounds. A recent study compared a large number of Canadian and Chinese toddlers.

Muscles as effectors

Edexcel   4.7.2–4

The body has a number of different effectors, but for most responses the effector is a muscle. There are three types of muscle in the body:

- skeletal/striated or voluntary muscle
- visceral/smooth or involuntary muscle
- cardiac muscle.

Smooth muscle is controlled by the autonomic nervous system, but skeletal muscle is controlled by motor neurones of the somatic nervous system.

How do motor neurones control muscle tissue?

The link to muscle tissue is by **motor end plates** which have close proximity to the sarcoplasm of the muscle tissue. The motor end plates have a greater surface area than a synaptic knob, but their action is very similar to the synaptic transmission described on page 31. Action potentials result in muscle contraction.

No contraction would take place without the acetylcholine transmitter being released from the motor end plate. When the sarcolemma (membrane) reaches the threshold level, then the action potential is conducted throughout the sarcolemma. Contraction is initiated!

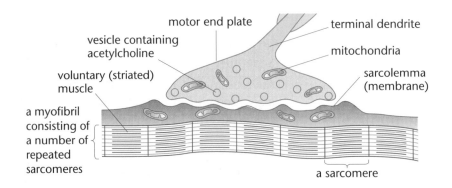

Skeletal muscle is also known as striated or striped muscle. The structure of a single muscle unit, the sarcomere, shows the striped nature of the muscle.

The sarcomere

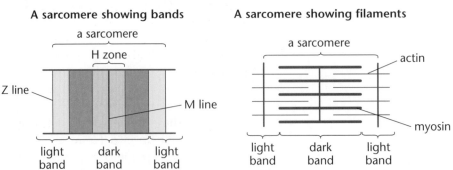

- The sarcomere consists of different filaments, thin ones (**actin**) and thick ones (**myosin**).
- These filaments form bands of different shades:
  - light band (I bands) – just actin filaments
  - dark band (A bands) – just myosin filaments or myosin plus actin.

- During contraction, the filaments slide together to form a shorter sarcomere.
- As this pattern of contraction is repeated through 1000s of sarcomeres, the whole muscle contracts.
- Actin and myosin filaments slide together because of the formation of cross bridges which alternatively build and break during contraction.
- Cross bridge formation is known as the 'ratchet mechanism'.

How does the 'ratchet mechanism' work?

To answer this question, the properties of actin and myosin need to be considered. The diagram below represents an actin filament next to a myosin filament. Many 'bulbous heads' are located along the myosin filaments (just one is shown). Each points towards an actin filament.

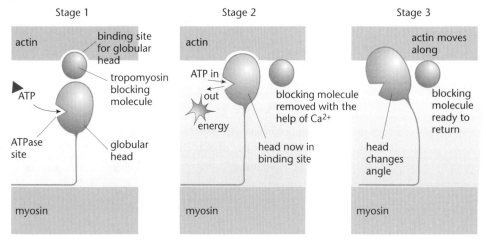

The sequence of the ratchet mechanism

- Once an action potential is generated in the muscle tissue then Ca^{2+} ions are released from the **reticulum**, a structure in the **sarcoplasm**.
- Part of the **globular head** of the myosin has an **ATPase** (enzyme) site.
- Ca^{2+} ions activate the myosin head so that this ATPase site hydrolyses an ATP molecule, **releasing energy**.
- Ca^{2+} ions also bind to **troponin** in the actin filaments, this in turn removes **blocking molecules (tropomyosin)** from the actin filament.
- This exposes the **binding sites** on the **actin** filaments.
- The globular heads of the myosin then bind to the newly exposed sites forming **actin–myosin cross bridges**.
- At the stage of energy release the myosin **heads change angle**.
- This change of angle moves the actin filaments towards the centre of each sarcomere and is termed the **power stroke**.
- More ATP binds to the myosin head, effectively causing the cross bridge to become straight and the tropomyosin molecules once again block the actin binding sites.
- The myosin is now 'cocked' and ready to repeat the above process.
- Repeated cross bridge formation and breakage results in a rowing action shortening the sarcomere as the filaments slide past each other.

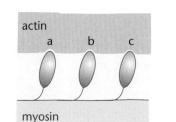 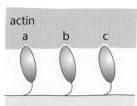 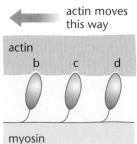

How does skeletal muscle produce movement?

Motor neurones control the skeletal muscle via motor end plates. The skeletal muscles move the bones via their tendon attachments. The muscles work in **antagonistic** pairs, i.e. **opposing** each other. In the arm, when the biceps contracts the forearm is lifted. At the same time the triceps relaxes. If the forearm is to be lowered then the triceps contracts and the biceps relaxes.

Types of skeletal muscle fibres

Under the microscope, skeletal muscle fibres can be divided into two types. They are slow and fast fibres. They do different jobs and so are found in different muscles.

Fibre type	Slow fibres	Fast fibres
Speed of contraction	Slow / tonic	Fast / twitch
Size of motor neurone	Small	Large
Activity used for	Responses such as maintaining posture	Quick responses such as jumping
Duration of use	Hours	Less than 5 minutes
Power produced	Low	High
Number of mitochondria	Many	Few
Found	In muscles such as the buttocks	In muscles such as the biceps
Myoglobin content	High so red	Low so white

2.5 Plant sensitivity

After studying this section you should be able to:

- *understand the range of tropisms which affect plant growth*
- *understand how auxins, gibberellins and cytokinins control plant growth*
- *understand how phytochromes control the onset of flowering in plants*

LEARNING SUMMARY

Plant growth regulators

Edexcel 5.8.2/8

External stimuli such as light can affect the direction of plant growth. A **tropism** is a **growth response** to an external stimulus. It is important that a plant grows in a direction which will enable it to obtain maximum supplies. Plant regulators are substances produced in minute quantities and tend to interact in their effects.

> Growth response to light is **phototropism**
> Growth response to gravity is **geotropism**
> Growth response to water is **hydrotropism**
> Growth response to contact is **thigmotropism**
>
> Tropisms can be positive (towards) or negative (away from).
>
> KEY POINT

Phototropism

This response is dependent upon the stimulus – light affecting the growth regulator, **auxin** (indoleacetic acid).

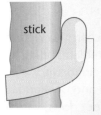

stick

auxin high
concentration here
so cells elongate

Thigmotropism helps a
climbing plant like the
runner bean to grow in
a twisting pattern
around a stick. Auxin is
redistributed away from
the contact point so the
outer cells elongate
giving a stronger outer
growth.

The diagrams show
tropic responses to light
and auxin.

Auxin and growing shoots

Auxin is produced by cells undergoing mitosis, e.g. growing tips. If a plant shoot is illuminated from one side then the auxin is redistributed to the side furthest from the light. This side grows more strongly, owing to the elongation of the cells, resulting in a bend towards the light. The plant benefits from increased light for photosynthesis. Up to a certain concentration, the degree of bending is proportional to auxin concentration.

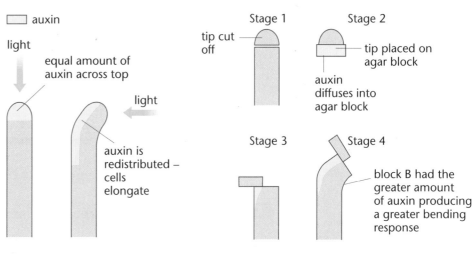

Tropisms in response to light from different directions

Tropism in response to auxin

Auxin research

Many investigations of auxins have taken place using the growing tips of grasses. Where a growing tip is removed and placed on an agar block, auxin will diffuse into the agar. Returning the block to the mitotic area stimulates increased cell elongation to the cells receiving a greater supply of auxin.

Is the concentration of auxin important?

It is important to consider the implications of the concentration of auxin in a tissue. The graph below shows that at different concentrations, auxin affects the shoot and the root in different ways.

Analyse this graph
carefully. It shows how
the same substance can
both stimulate or inhibit,
depending on
concentration.

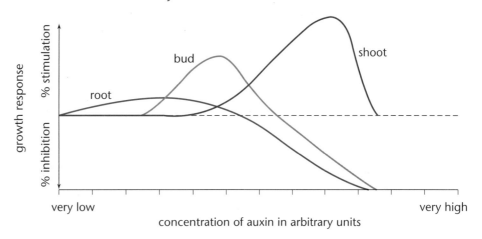

The graph shows:

- auxin has no effect on a shoot at very low concentration
- at these very low concentrations root cell elongation is stimulated
- at higher concentrations the elongation of shoot cells is stimulated
- at these higher concentrations auxin inhibits the elongation of root cells.

Auxin and root growth

The graph on page 50 shows that auxin affects root cells in a different way at different concentrations. At the root tip, auxin accumulates at a lower point because of gravity. This inhibits the lower cells from elongating. However, the higher cells at the tip have a low concentration of auxin and do elongate. The net effect is for the stronger upper cell growth to bend the root downwards. The plant therefore has more chance of obtaining more water and mineral ions.

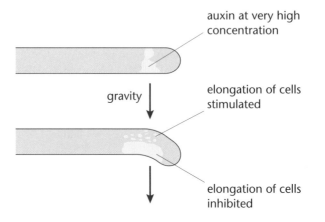

Plant growth regulators

In your examination, look out for data which will be supplied, e.g. the growth regulator gibberellin may be linked to falling starch levels in a seed endosperm and increase in maltose. Gibberellin has stimulated the enzymic activity.

Hormone	Some key functions
auxin	increased cell elongation, suppression of lateral bud development
gibberellin	cell elongation, ends dormancy in buds, promotes germination of seeds by activating hydrolytic enzymes such as amylase (food stores are mobilised!)
cytokinin	increased cell division, increased cell enlargement in leaves
ethene	promotes ripening of food

Phytochrome

Phytochrome:

- is a regulatory substance
- is photosensitive
- has a number of different roles
- controls the onset of flowering in plants
- exists in two different forms – **phytochrome red (P_R)** and **phytochrome far red (P_{FR})**.

The two forms are inter-convertible as shown below.

The terms P_R and P_{FR} refer to the peak wavelengths of light, absorbed by each substance:

P_R absorbs a peak of 665 nm

P_{FR} absorbs a peak of 725 nm.

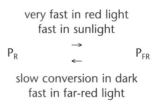

The above inter-conversion of the phytochromes is part of the mechanism that controls the onset of flowering.

Different species of plants respond to different day lengths during the year. Specific day length triggers the development of the flower buds.

This is known as **photoperiodism**.

There are three categories of plant:

- long-day plants, e.g. petunias (**need P$_{FR}$ to flower**)
- short-day plants, e.g. chrysanthemums (**need P$_R$ to flower**)
- day-neutral plants, e.g. tomatoes.

The day-length and night-length bars below show the proportion of light and dark and the effect on the flowering.

long-day plants in summer

- P$_R$ ⟶ P$_{FR}$ (fast conversion in light)
- flowering promoted (long-day plants need P$_{FR}$)

short-day plants in summer

- P$_R$ ⟶ P$_{FR}$
- flowering NOT promoted (short-day plants need P$_R$)

long-day plants in winter

- P$_{FR}$ ⟶ P$_R$ (slow process but the night is long enough)
- flowering not promoted (long-day plants need P$_{FR}$)

short-day plants in summer

- P$_{FR}$ ⟶ P$_R$ (slow process but long enough at night to make P$_R$)
- flowering promoted (short-day plants need P$_R$)

Key

◼ darkness

☐ sunlight

In your examinations, look out for data about day length. Particularly look for the short-day data where a flash of light occurs during a dark period. This is enough to make P$_{FR}$ which will stimulate long-day plants to flower. This is due to the rapid P$_R$ → P$_{FR}$ process.

Question 1

(a) The diagram shows a neurone.

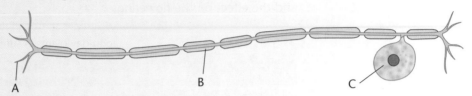

(i) What type of neurone is shown in the diagram? [1]

sensory neurone

(ii) What structure would be found at A? [1]

a receptor

(iii) Where precisely in the body is structure C found? [1]

in the dorsal root ganglion

(iv) Structure B is covered by a myelin sheath.
Explain the function of the myelin sheath. [3]

The myelin sheath insulates the neurone;

Ions can only pass into the neurone at the nodes of Ranvier;

This speeds up the transmission of the nerve impulse.

(b) The diagram shows a cone from the retina.

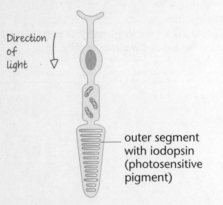

Direction of light

outer segment with iodopsin (photosensitive pigment)

(i) Place an arrow on the diagram to show the direction in which light reaches the cone. [1]

(ii) What is the function of the iodopsin in the outer segment? [4]

when stimulated by light of the correct wavelength – breaks down to release opsin; opsin opens ion channels in the membranes; this can lead to the generation of an action potential in a bipolar cell.

(iii) How do cones contribute to high visual acuity? [3]

cones are tightly packed giving a high surface area; each cone synapses onto a single bipolar neurone; so the greater detail gives higher resolution.

Practice examination questions

1   The growing tips were removed from oat stems. Agar blocks containing different
    concentrations of synthetic auxin (IAA) replaced the tips on the oat stems. The
    plants were allowed to grow for a period then the angle of curvature of the stems
    was measured. The results are shown in the graph below.

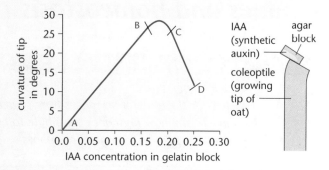

(a) What is the relationship between IAA concentration and curvature of the stem
    between points:

(i) A and B   [1]
(ii) C and D?   [1]

(b) Explain how IAA causes a curvature in the oat stems.   [2]

(c) Explain the effect a much higher concentration of IAA would have on the
    curvature of oat stems.   [2]

                                                                       [Total: 6]

2   The diagram below shows a single sarcomere just before contraction.

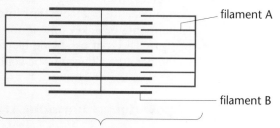

a sarcomere

(a) Name filaments A and B.   [2]

(b) What stimulus causes the immediate contraction of a sarcomere?   [1]

(c) What happens to each type of filament during contraction?   [2]

                                                                       [Total: 5]

3   The diagram below shows the profile of an action potential.

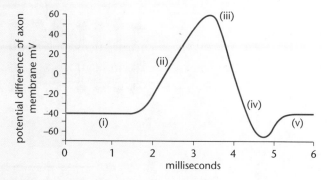

Explain what happens in the axon at each stage shown on the diagram.   [10]

                                                                      [Total: 10]

Chapter 3

Homeostasis

The following topics are covered in this chapter:

- Hormones and homeostasis
- Homeostasis and exercise
- Temperature control in a mammal

3.1 Hormones and homeostasis

After studying this section you should be able to:

- *define homeostasis*
- *describe the route of hormones from source to target organ*
- *understand how hormones contribute to homeostasis*
- *recall the roles of a range of hormones*

LEARNING SUMMARY

The endocrine system

Edexcel   5.7.15

The endocrine system secretes a number of chemicals known as **hormones**. Each hormone is a substance produced by an **endocrine gland**, e.g. adrenal glands produce the hormone adrenaline. Each hormone is **transported in the blood** and has a **target organ**. Once the target organ is reached, the hormone **triggers a response** in the organ. Many hormones do this by **activating enzymes**. Others **activate genes**, e.g. steroids.

> The great advantage of homeostasis is that the conditions in the environment fluctuate but conditions in the organism remain stable.

> The endocrine and nervous systems both contribute to **coordination** in animals. They help to regulate internal processes. **Homeostasis** is the maintenance of a **constant internal environment**. Nerves and hormones have key roles in the maintenance of this **steady internal state**. Levels of pH, blood glucose, oxygen, carbon dioxide and temperature all need to be controlled.
>
> **KEY POINT**

How does a hormone trigger a cell in a target organ?

Hormones are much slower in eliciting a response than the nervous system. Rather than having an effect in milliseconds like nerves, hormones take longer. However, effects in response to hormones are often **long lasting**.

The diagram below shows one mechanism by which hormones activate target cells.

> Did you know?
> Each enzyme shown is constantly re-used as an active site is left free.

> Look carefully at this mechanism! Just **ONE hormone molecule** arriving at the cell releases an enzyme which can be used **many** times. In turn, another enzyme is produced which can be used **many** times. One hormone molecule leads to **amplification**. This is a cascade effect.

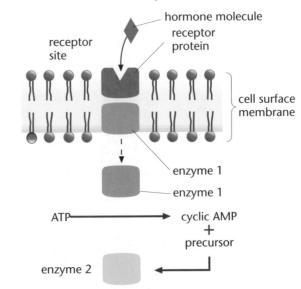

Hormones that are proteins work in this way because they are unable to enter the cell. Steroid hormones, (e.g. oestrogen) can pass through the cell membrane as they are lipid soluble. They often work by switching genes on, which will simulate protein synthesis.

3.2 Homeostasis and exercise

After studying this section you should be able to:

- *understand how the heart rate is changed in response to exercise*
- *understand how the breathing rate is changed in response to exercise*

LEARNING SUMMARY

Control of the heart rate

Edexcel 5.7.12-5.7.13

Read p.49 to remind yourself how the SA node controls the heart beat.

The nervous system and the endocrine system work together in the body to achieve homeostasis. An example of this cooperation is in the control of the heart rate. Cardiac muscle in the heart is **myogenic**. This means that it will contract without a nerve impulse. The rate is set by the pacemaker in the SA node but this can be adjusted by the **cardiovascular centre in the medulla**. This allows the body to match the output of the heart to the demands of exercise.

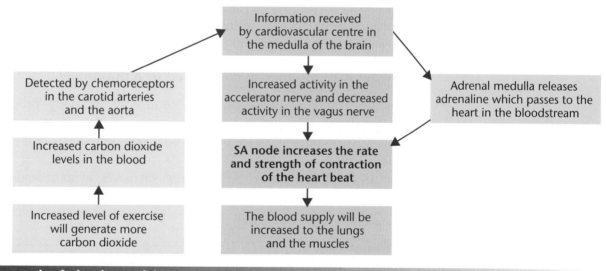

Information received by cardiovascular centre in the medulla of the brain

Detected by chemoreceptors in the carotid arteries and the aorta

Increased activity in the accelerator nerve and decreased activity in the vagus nerve

Adrenal medulla releases adrenaline which passes to the heart in the bloodstream

Increased carbon dioxide levels in the blood

SA node increases the rate and strength of contraction of the heart beat

Increased level of exercise will generate more carbon dioxide

The blood supply will be increased to the lungs and the muscles

Control of the breathing rate

Edexcel 5.7.13

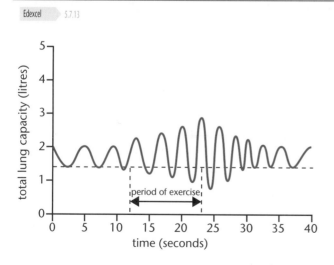

There is little point in the heart beating faster during exercise if the breathing rate is not increased as well. This is controlled by the **ventilation centre in the medulla**. The same chemoreceptors that trigger the increase in heart rate also send impulses to the ventilation centre. This will increase the rate at which impulses are sent to the intercostal muscles and diaphragm. This will increase the rate and depth of breathing, allowing increased gaseous exchange in the lungs.

Changes in rate and depth of breathing can be measured using a machine called a spirometer. It produces a trace which can be used to measure the breathing rate and the tidal volume.

The tidal volume is the volume of air exchanged per breath during quiet breathing. This will increase during exercise.

KEY POINT

3.3 Temperature control in a mammal

After studying this section you should be able to:

- outline the processes which contribute to temperature regulation in a mammal
- understand how nervous and endocrine systems work together to regulate body temperature
- understand how internal processes are regulated by negative feedback

LEARNING SUMMARY

What are the advantages of controlling body temperature?

Edexcel 5.7.16

It is advantageous to maintain a constant body temperature so that the enzymes which drive the processes of life can function at an optimum level.

- **Endothermic** (warm blooded) animals can maintain their core temperature at an optimal level. This allows internal processes to be consistent. The level of activity of an endotherm is likely to fluctuate less than an ectotherm.
- **Ectothermic** (cold blooded) animals have a body temperature which fluctuates with the environmental temperature. As a result there are times when an animal may be vulnerable due to the enzyme-driven reactions being slow. When a crocodile (ectotherm) is in cold conditions, its speed of attack would be slow. When in warm conditions, the attack would be rapid.

How is temperature controlled in a mammal?

The key structure in homeostatic control of all body processes is the **hypothalamus**. The regulation of temperature involves thermoreceptors in the skin, body core and blood vessels supplying the brain, which link to the hypothalamus.

The diagram below shows how the peripheral nerves, hypothalamus and pituitary gland integrate nervous and endocrine glands to regulate temperature.

> Once the blood temperature decreases, the heat gain centre of the hypothalamus is stimulated. This leads to a rise in blood temperature which, in turn, results in the heat loss centre being stimulated. This is negative feedback! The combination of the two, in both directions, contributes to homeostasis.

> The **hypothalamus** has **many functions**. It controls thirst, hunger, sleep and it stimulates the production of many hormones other than those required for temperature regulation.

> If there is an increase in core temperature then the hypothalamus stimulates greater heat loss by:
> - vasodilation (dilation of the skin arterioles)
> - relaxing of erector-pili muscles so that hairs lie flat
> - more sweating
> - behavioural response in humans to change to thinner clothing.

Temperature regulation model

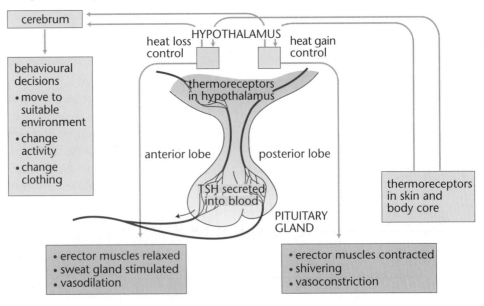

When the hypothalamus receives sensory information **heat loss** or **heat gain** control results.

A capillary bed

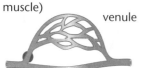

arteriole
(a sphincter
muscle)

venule

shunt vessel

artery

vein

A **fall** in temperature results in the following control responses.

Vasoconstriction
- Arteriole control is initiated by the hypothalamus which results in efferent neurones stimulating constriction of the arteriole sphincters of skin capillary beds.
- This deviates blood to the core of the body, so less heat energy is lost from the skin.

Contraction of the erector-pili muscles
- Erector-pili muscle contraction is initiated in the hypothalamus being controlled via efferent neurones.
- Hairs on skin stand on end and trap an insulating layer of air, so less heat energy is lost from the skin.

Sweat reduction
- The sweat glands control is also initiated in the hypothalamus, and is controlled via efferent neurones.
- Less heat energy is lost from the skin by evaporation of sweat.

Shivering
- Increased muscular contraction is accompanied by heat energy release.

Behavioural response
- A link from the hypothalamus to the cerebrum elicits this voluntary response.
- This could be to switch on the heating, put on warmer clothes, etc.

Increased metabolic rate
- The hypothalamus produces a release factor substance.
- This stimulates the anterior part of the pituitary gland to secrete TSH (thyroid stimulating hormone).
- TSH reaches the thyroid via the blood.
- Thyroid gland is stimulated to secrete thyroxine.
- Thyroxine increases respiration in the tissues increasing the body temperature.

Once a higher thyroxine level is detected in the blood the release factor in hypothalamus is inhibited so TSH release by the pituitary gland is prevented. This is **negative feedback**.

An **increase** in body temperature results in almost the **opposite** of each response described for a fall of temperature.

Vasodilation
- Arterioles of capillary beds dilate allowing more blood to skin capillary beds.

Relaxation of erector-pili muscles
- Hairs lie flat, no insulating layer of air trapped.
- More heat loss of skin.

Sweat increase
- More sweat is excreted so more heat energy from the body is needed to evaporate the sweat, so the organism cools down.

Behavioural response
- This could be to move into the shade or consume a cold drink.

Note
(a) the outline for heat loss methods does not show the nerve connections. Efferent neurones are again coordinated via the hypothalamus!
(b) heat is lost from the skin via a combination of **conduction**, **convection** and **radiation**.

Progress check

Hormone X stimulates the production of a substance in a cell of a target organ. The following statements outline events which result in the production of the substance but are in the wrong order. Write the correct order of letters.

A Hormone X is transported in the blood.
B Hormone X binds with a receptor protein in the cell surface membrane.
C The enzyme catalyses a reaction, forming a product.
D Hormone X is secreted by a gland.
E This releases an enzyme from the cell surface membrane.

D, A, B, E, C.

Sample question and model answer

The graph shows a person's breathing measured using a spirometer. The person breathed quietly for one minute then breathed out and in as deeply as they could.

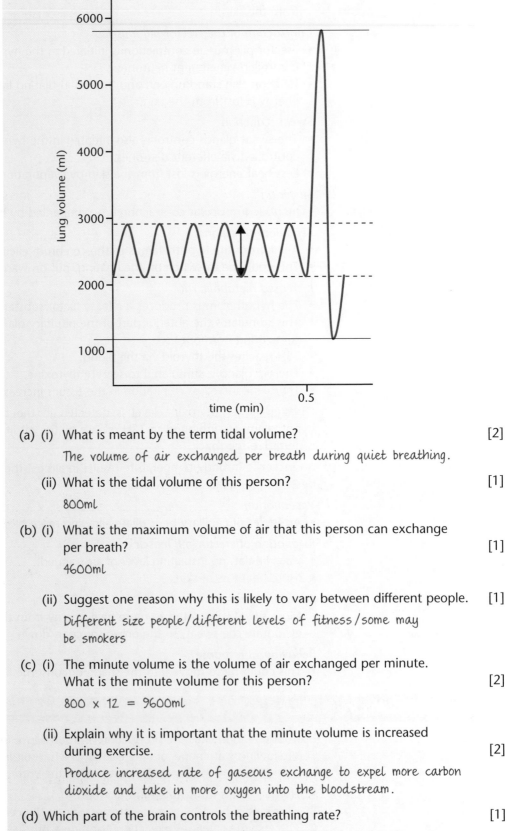

(a) (i) What is meant by the term tidal volume? [2]

The volume of air exchanged per breath during quiet breathing.

(ii) What is the tidal volume of this person? [1]

800ml

(b) (i) What is the maximum volume of air that this person can exchange per breath? [1]

4600ml

(ii) Suggest one reason why this is likely to vary between different people. [1]

Different size people / different levels of fitness / some may be smokers

(c) (i) The minute volume is the volume of air exchanged per minute. What is the minute volume for this person? [2]

800 x 12 = 9600ml

(ii) Explain why it is important that the minute volume is increased during exercise. [2]

Produce increased rate of gaseous exchange to expel more carbon dioxide and take in more oxygen into the bloodstream.

(d) Which part of the brain controls the breathing rate? [1]

The respiratory centre in the medulla.

Practice examination questions

1 (a) Complete the table below to compare the nervous and endocrine systems. Put a tick in each correct box for the features shown.

	Nervous system	Endocrine system
Usually have longer lasting effects		
Have cells which secrete transmitter molecules		
Cells communicate by substances in the blood plasma		
Use chemicals which bind to receptor sites in cell surface proteins		
Involve the use of Na^+ and K^+ pumps		

[2]

(b) Name the process which keeps the human body temperature and water content of blood regulated. [1]

[Total: 3]

2 A person is exercising using an exercise bike.

Changes in his body are shown in the table.
The stroke volume is the volume of blood pumped out by the heart per beat.
The cardiac output is the volume of blood pumped out by the heart in one minute.

Work rate in watts	Heart rate in beats per minute	Stroke volume in cm^3	Cardiac output in cm^3
0		65	5590
60	106	84	8904
100	122		14640
140	143	130	18590
160	156	125	18500

(a) Work out the person's heart rate at rest and their stroke volume when working at 100 watts. [2]

(b) Suggest why these studies are usually performed using an exercise bike rather than by getting a person to run. [1]

(c) Describe the changes that occur in the stroke volume during the experiment. [3]

(d) Outline the changes that occur in the body that lead to the increase in heart rate during the experiment. [5]

[Total: 11]

Chapter 4

Genetics, gene technology and selection

The following topics are covered in this chapter:

- *Genes and protein synthesis*
- *Mapping and manipulating genes*
- *Selection and speciation*

4.1 Genes and protein synthesis

After studying this section you should be able to:

- *explain how proteins are produced in cells*
- *describe various methods by which gene expression is controlled*

LEARNING SUMMARY

Genes and protein synthesis

Edexcel ▶ 4.6.2–3

A gene is a section of DNA which controls the production of a polypeptide in an organism. The total effects of all of the genes of an organism are responsible for the characteristics of that organism. Each protein contributes to these characteristics whatever its role, e.g. structural, enzymic or hormonal.

The order of bases in the gene is called the genetic code and will code for the order of amino acids in the polypeptide. The order of amino acids is called the primary structure of the protein and will determine how the protein folds up to form the secondary and tertiary structures. The formation of a protein molecule is called protein synthesis.

Protein synthesis

The process of protein synthesis involves the DNA and several other molecules.

- **Messenger RNA:** This is a single-stranded nucleotide chain that is made in the nucleus. It carries the complementary DNA code out of the nucleus to the ribosomes in the cytoplasm.
- **mRNA polymerase:** This is the enzyme that joins the mRNA nucleotides together to form a chain.
- **ATP:** This is needed to provide the energy to make the mRNA molecule and to join the amino acids together.
- **tRNA:** This is a short length of RNA that is shaped rather like a clover leaf. There is one type of tRNA molecule for every different amino acid. The tRNA molecule has three unpaired bases that can bind with mRNA on one end and a binding site for a specific amino acid on the other end.

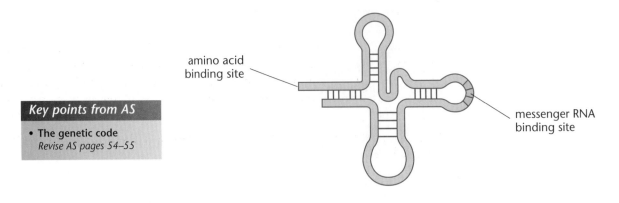

amino acid
binding site

messenger RNA
binding site

Key points from AS

- **The genetic code**
  Revise AS pages 54–55

The following diagrams show protein synthesis.

1 In the nucleus **RNA polymerase** links to a start code along a DNA strand.

2 RNA polymerase moves along the DNA. For every organic base it meets along the DNA a complementary base is linked to form mRNA (**messenger RNA**).

> There is no thymine in mRNA. Instead there is another base, uracil.

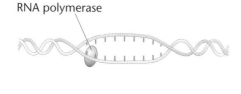

Pairing of organic bases			
DNA	G	C	T	A
mRNA	C	G	A	U

3 RNA polymerase links to a stop code along the DNA and finally the mRNA **moves to a ribosome**. The DNA stays in the nucleus for the next time it is needed.

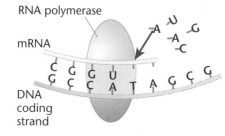

The transfer of the code from DNA to mRNA is called transcription.

4 Every three bases along the mRNA make up one **codon** which codes for a specific amino acid. Three complementary bases form an **anticodon** attached to one end of tRNA (**transfer RNA**). At the other end of the RNA is a specific amino acid.

5 All along the mRNA the tRNA 'partner' molecules enable each amino acid to bond to the next. A chain of amino acids (**polypeptide**) is made, ready for release into the cell.

> Note the link between each pair of amino acids along a polypeptide – the peptide link.

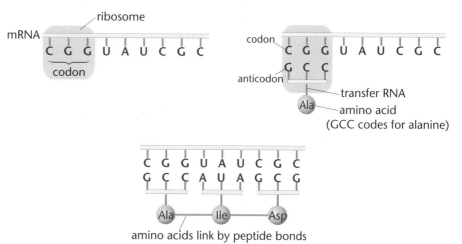

amino acids link by peptide bonds

The conversion of the mRNA code to a sequence of amino acids is called translation.

Control of protein synthesis

Edexcel   4.6.4

In a multicellular organism, every cell contains all the genetic material needed to make every protein that the organism requires. However, as they develop, cells become specialised. This means that they do not need to use every protein and so it would be a waste to make every protein all the time. Genes must be switched on and switched off.

Most of the early work on gene regulation was carried out on bacteria which are prokaryotic.

Jacob and Monod's theory

During the 1950s, Jacob and Monod found that the bacterium *E.coli* would only produce the enzyme lactase if lactose was present in the growth medium. The production of lactose was controlled by three different genes:

- a structural gene codes for the enzyme
- an operator gene which turns the structural gene on
- a regulator gene that produces a chemical that usually stops the action of the operator gene.

If lactose is present, the action of the chemical inhibitor is blocked and lactase is made.

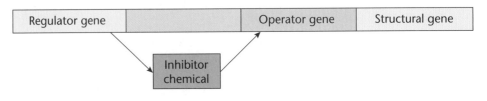

> **KEY POINT**
>
> The combination of the three genes controlling lactase production is called the *lac* operon.

Gene control in eukaryotes

In mature plants, many cells remain totipotent but in mature animals these totipotent or **stem cells** are harder to find.

In eukaryotic cells, gene regulation seems to be much more complicated. Cells in the early embryo are called **totipotent**. This means that they can develop into any type of cell. These cells produce all the cells of a multicellular organism and the specialised cells have to be produced in the correct place.

Scientists are trying to work out how this is done and have found genes called **homeobox genes**. These genes seem to produce proteins that act as transcription factors turning on other genes. Similar homeobox genes have been found in animals, plants and fungi.

Other factors from the cytoplasm can also effect transcription. **Steroid hormones** such as oestrogen can bind with **receptors** in the cytoplasm and then move into the nucleus causing genes to be transcribed.

There is much interest at present in the possible use of siRNA to treat various genetic conditions.

Scientists have recently found a different type of RNA. This is a small double-stranded molecule called siRNA (small interfering RNA). This seems to silence the action of certain genes.

Post transcriptional modification

It is now known that less than 98% of our DNA actually codes for protein. The rest is often called 'junk DNA'. However, much of this junk DNA may be important in regulating (such as coding for siRNA).

Each gene contains some lengths of DNA that code for the protein, called **exons**. Scattered in between the exons are **introns** that do not code for the protein.

☐ = start or stop codons

☐ = exons

☐ = introns

When mRNA is made by transcription, the entire DNA for that gene is transcribed. The mRNA that corresponds to the introns is then removed. This is called **post transcriptional modification**.

4.2 Mapping and manipulating genes

After studying this section you should be able to:

- *describe a range of techniques used in mapping genes*
- *describe how genes can be transferred between a range of different organisms*

Identifying genes

Edexcel 4.6.5–7, 5.8.19

In order to identify the genes of an individual, a number of different processes are used.

Polymerase chain reaction

The polymerase chain reaction (PCR) is used to make numerous copies of a section of DNA. This is called **amplifying** the DNA. It uses the principle of semiconservative replication of DNA to produce new molecules that can in turn act as templates to produce more molecules. This therefore sets up a chain reaction. The enzyme DNA polymerase is used to copy the DNA.

Electrophoresis

This is used to separate sections of DNA according to their size.

Enzymes called **restriction endonucleases** can be used to cut up an organism's DNA.

- DNA sections are put into a well in a slab of agar gel.
- The gel and DNA are covered with buffer solution which conducts electricity.
- Electrodes apply an electrical field.
- Phosphate groups on DNA are negatively charged causing DNA to move towards the anode.
- Smaller pieces of DNA move more quickly along the agar track; larger ones move more slowly, leading to the formation of bands.

Genetic fingerprinting

PCR and electrophoresis have many applications. DNA is highly specific so the bands produced using this process can help with identification. In some crimes, DNA is left at the scene. Blood and semen both contain DNA specific to an individual. DNA evidence can be checked against samples from suspects. This is known as **genetic fingerprinting**. Genetic fingerprinting can be used in paternity disputes. Each band of the DNA of the child must correspond with a band from *either* the father or the mother.

Isolating genes

Along chromosomes are large numbers of genes. Scientists may need to identify and isolate a useful gene; one way of doing this is to use the enzyme **reverse transcriptase**. This is produced by viruses known as **retroviruses**. Reverse transcriptase has the ability to help make DNA from mRNA. This allows copies of eukaryotic genes to be made without the junk DNA or introns.

Stage 1

When a polypeptide is about to be made at a mRNA ribosome, reverse transcriptase allows a strand of its coding DNA to be made. This strand has had the introns removed.

Stage 2

The single stranded DNA is parted from the mRNA.

Stage 3

The other strand of DNA is assembled using DNA polymerase.

Using this principle, the exact piece of DNA which codes for the production of a vital protein can be made, without the introns.

Progress check

1 A length of DNA was prepared and then electrophoresis was used to separate the sections. The statements below describe the process of electrophoresis but they are in the wrong order. Write the letters in the correct sequence.

A electrodes apply an electrical field

B DNA sections are put into a well in a slab of agar gel

C smaller pieces of DNA move more quickly along the agar track with larger ones further behind

D the gel and DNA are then covered with buffer solution which conducts electricity

E restriction endonucleases can be used to cut up the DNA

2 Reverse transcriptase is an enzyme which enables the production of DNA from RNA. Work out the sequence of organic bases along a DNA strand made from the following RNA molecule.

A A U G C C C G G A U U

2 TTACGGGCCTAA
1 E B D A C

The Human Genome Project

The Human Genome Project is an analysis of the complete human genetic make-up which has mapped the organic base sequences of the nucleotides along our DNA.

A brief history

- 1977 Sanger devised DNA base sequencing.
- 1986 The Human Genome Project was initiated in the USA and the UK.
- 1996 30 000 genes were mapped.
- 1999 one billion bases were mapped including all of chromosome 22.
- 2000 chromosome 21 was mapped with the human genome almost complete.
- 2001 human genome mapping complete.

Effects of single nucleotide polymorphism

Example

5 base sequences from five people →

GTATAGCCGCAT	1
GTATAGCCGCAT	1
GTATAGCCGCAT	1
GTATAGCCGCCT	2
GTATAGCCGCCT	2

Version 1 = ●
Version 2 = ●

Proportion of the SNP in healthy members of population:

Proportion of the SNP in diseased members of population:

A greater incidence of an SNP in people with a disease may point to a cause.

Single nucleotide polymorphisms (SNPs)

Around 99.9% of human DNA is the same in all individuals. Merely 0.1% is different! The different sequences in individuals can be the result of **single nucleotide polymorphism**. One base difference from one individual to another at a site may have no difference. Up to a maximum of six different codons can code for one amino acid. An SNP will not necessarily have any effect.

Some SNPs do change a protein significantly. Such changes may result in genetic disease, resistance or susceptibility to disease.

How can the mapping of SNPs be useful?

- The mapping of SNPs along chromosomes signpost where base differences exist.
- Across the gene pool a pattern of SNP positions will be evident.
- There may be a high frequency of common SNPs found in the DNA of people with a specific disease.
- This highlights interesting sites for future research and will help to find answers to genetic problems.

Benefits obtained from the Human Genome Project

Ultimately, the human genome data will be instrumental in the development of drugs to treat genetic disease. Additionally, by analysis of parental DNA, it will be possible to give the probability of the development of a specific disease or susceptibility to it, in offspring. Fetal DNA, obtained through amniocentesis or by chorionic villi sampling, will give genetic information about an individual child.

Genetic counsellors will have more information about an individual than ever before. Companies will be able to produce 'designer drugs' to alleviate the problems which originate in our DNA molecules. Soon the race will begin to produce the first crop of drugs to treat or even cure serious genetic diseases. Look to the media for progress updates.

Manipulating DNA

Edexcel   5.8.20

Scientists have developed methods of manipulating DNA. It can be transferred from one organism to another. Organisms which receive the DNA then have the ability to produce a new protein. This is one example of **genetic engineering**.

> The genetic code is universal. This means that it is possible to move genes from one organism to another and the recipient organism may be from a different species. The DNA will still code for the same protein. **KEY POINT**

Genes have now been transferred to and from many different types of organisms.

Here are some examples:

- From humans to bacteria: this technique produced the first commercially available genetically engineered product, insulin.
- From plants to plants: this technique has been used to produce GM crops such as Golden Rice that contain vitamin A.
- Into humans: this technique may be successful in treating genetic conditions such as cystic fibrosis.

Gene transfer to bacteria

The gene which produces human insulin was transferred from a human cell to a bacterium. The new microorganism is known as a **transgenic bacterium**. The process which follows shows how a human gene can be inserted into a bacterium.

The human gene for insulin is produced using an enzyme called **reverse transcriptase**. This converts the mRNA coding for insulin back into DNA. In this way all the introns are removed. Then the DNA can be inserted.

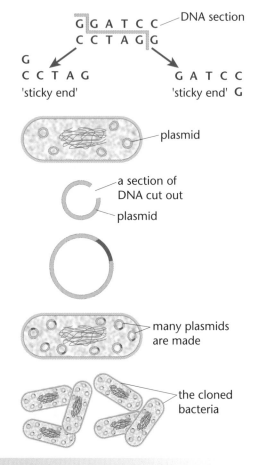

1 An enzyme known as **restriction endonuclease** cuts the DNA and the gene was removed. Each time a cut was made the two ends produced were known as 'sticky ends'.

G G A T C C — DNA section
C C T A G G

G
C C T A G   G A T C C
'sticky end'   'sticky end' G

Restriction endonucleases are produced by some bacteria as a defence mechanism. They cut up the DNA of invading viruses. This can be exploited during gene transfer.

2 Circles of DNA called **plasmids** are found in bacteria.

plasmid

*Note that **both** the donor DNA and recipient plasmid DNA are cut with the same enzyme. This allows the new gene to be a matching fit.*

3 A plasmid was taken from a bacterium and cut with the same restriction endonuclease.

a section of DNA cut out
plasmid

4 The human gene was inserted into the plasmid. It was made to fix into the open plasmid by another enzyme known as **ligase**.

Many exam candidates fail to state that the plasmids are cloned inside the bacterium.

5 The plasmid **replicated** inside the bacterium.

many plasmids are made

The bacteria themselves are also cloned. There may be two marks in a question for each cloning point!

6 Large numbers of the new bacteria were produced. Each was able to secrete perfect human insulin, helping diabetics all over the world.

the cloned bacteria

Gene transfer to plants

Inserting genes into crop plants is becoming increasingly important in meeting the needs of a rising world population. One example of this is the production of a type of rice called Golden Rice. This contains a gene that produces vitamin A. The aim is to prevent vitamin A deficiency which can lead to blindness.

In **plants** there is an important technique which uses a **vector** to insert a novel gene. The vector is the bacterium *Agrobacterium tumefaciens*.

Agrobacterium tumefaciens
- This is a **pathogenic bacterium** which **invades** plants forming a gall (abnormal growth).
- The bacterium contains **plasmids** (circles of DNA) which carry a gene that stimulates tumour formation in the plants it attacks.
- The part of the plasmid which does this is known as the **T-DNA region** and can insert into any of the chromosomes of a host plant cell.
- Part of the T-DNA controls the production of two growth hormones, auxin and cytokinin.
- The extra quantities of these hormones stimulate rapid cell division, the cause of the tumour.

KEY POINT

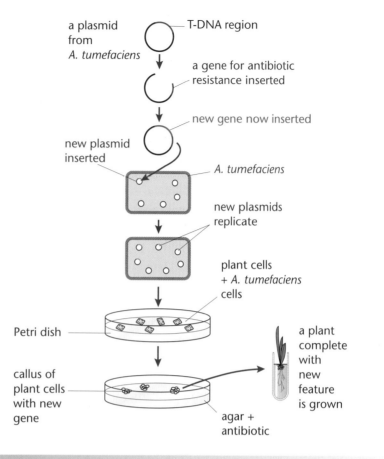

How can *Agrobacterium tumefaciens* be used in gene transfer?

The principle of using *A. tumefaciens* can be used in gene transfer to many different plants. Applications are at an early stage of development.

KEY POINT

- Firstly, the DNA section controlling auxin and cytokinin was deleted, tumours were not formed, and cells of the plant retained their normal characteristics.
- A gene which gave the bacterial cell **resistance to a specific antibiotic** was inserted into the T-DNA position.
- The **useful gene** (e.g. the gene for vitamin A) was **inserted into a plasmid.**
- Plant cells, minus cell walls, were removed and put into a Petri dish with nutrients and *A. tumefaciens*, which contained the engineered plasmids.
- The cells were **incubated** for several days, then transferred to another Petri dish containing nutrients plus the specific antibiotic.
- **Only plant cells with antibiotic resistance and the desired gene grew.**
- Any surviving cells grew into a callus, from which an adult plant formed, complete with the transferred gene.

Inserting genes into humans

The idea of changing a person's genes in order to cure genetic disease is called **gene therapy.**

There are two main possibilities:

- **Somatic cell therapy** in which the genes are inserted into the cells of the adult where they are needed.
- **Germ line gene therapy** involves changing the genes of the gametes or early embryo. This means that all the cells of the organism will contain the new gene.

4.3 Selection and speciation

After studying this section you should be able to:

- understand the process of natural selection
- predict population changes in terms of selective pressures
- understand a range of isolating mechanisms and how a new species can be formed
- understand the difference between allopatric and sympatric speciation

LEARNING SUMMARY

Natural selection

Edexcel 4.5.21–22

Throughout the biosphere, communities of organisms interact in a range of ecosystems. Darwin travelled across the world in his ship, the *Beagle*, observing organisms in their habitats. In 1858 Darwin published *On the Origin of Species*. In this book he gave his theory of **natural selection**.

The key features of this theory are that as organisms interact with their environment:

- individual organisms of populations are not identical, and can **vary in both genotypes and phenotypes**
- **some organisms survive** in their environment other organisms **die** before reproducing, effectively being **deleted from the gene pool**
- surviving organisms **go on to breed** and **pass on their genes** to their offspring
- this **increases the frequency of the advantageous genes** in the population.

Learn this theory carefully then apply it to the scenarios given in your examination. Candidates often identify that some organisms die and others survive, but few go on to predict the inheritance of advantageous genes and the consequence to the species.

Consider these factors

- Adverse conditions in the environment could make a species extinct, but a range of genotypes increases the chances of the species surviving.
- Different genotypes may be suited to a changing environment, say, as a result of global warming.
- A variant of different genotype, previously low in numbers, may thrive in a changed environment and increase in numbers.
- Where organisms are well suited to their environment they have adaptations which give this advantage.
- If other organisms have been selected against, then more resources are available for survivors.
- Breeding usually produces many more offspring than the mere replacement of parents.
- Resources are limited so that competition for food, shelter and breeding areas takes place. Only the fittest survive!

What is selective pressure?

Selective pressure is the term given to a factor which has a direct effect on the numbers of individuals in a population of organisms, for example:

'It is late summer and the days without rainfall have caused the grassland to be parched. There is little food this year.'

Here the **selective pressure** is a **lack of food** for the herbivores. Species which are **best adapted** to this habitat **compete** well for the limited resources and go on to survive. Within a species there is a further application of the selective pressure as weaker organisms perish and the strongest survive.

In this example, the fact that the numbers of herbivores decrease is *another* selective pressure. This time numbers of predators may decrease.

Mutations are random

New genes can appear in a species for the first time, due to a form of mutation. Over thousands of years, repeated natural selection takes place, resulting in superb adaptations to the environment.

Considering these examples, it is no wonder that candidates seem to consider that the organisms actively adapt to develop in these ways. They suggest that the organisms themselves have control to make active changes. **This is not so! There is no control, no active adaptation**.

- The Venus fly trap with its intricate leaf structures captures insects. The insects decompose, supplying minerals to the mineral deficient soil.
- Crown Imperial lilies (*Fritillaria*) produce colourful flowers, and a scent of stinking, decomposing flesh. Flies are attracted and help pollination.
- The bee orchid flower is so like a queen bee that a male will attempt mating.

> *New genes appear by CHANCE!*

 KEY POINT

Selective pressures and populations

To find out more about the effects that selective pressures can have, the **normal distribution** must be considered. The distribution below is illustrated with an example.

The mean value is at the peak. There are fewer tall and short individuals in this example. A taller plant intercepts light better than a shorter one.

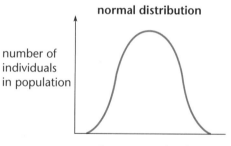

normal distribution

number of individuals in population

feature, e.g. height

The further distributions below show effects of selective pressures (shown by the blue arrows). Each is illustrated with an example.

Selective pressure at both ends of the distribution causes the extreme genotypes to die. This maintains the distribution around the mean value. Mean wing length is better for flight, better for prey capture.

Selective pressure results in death of slower animals. Many die out due to predators. Faster ones (with longer legs) pass on advantageous genes. Distribution moves to the right as the average individual is now faster.

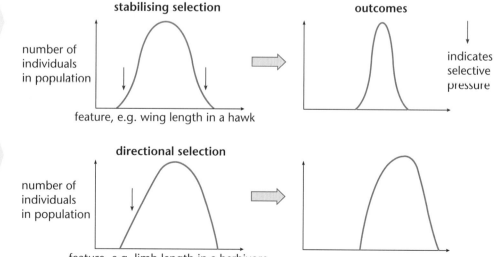

stabilising selection

number of individuals in population

feature, e.g. wing length in a hawk

outcomes

indicates selective pressure

directional selection

number of individuals in population

feature, e.g. limb length in a herbivore

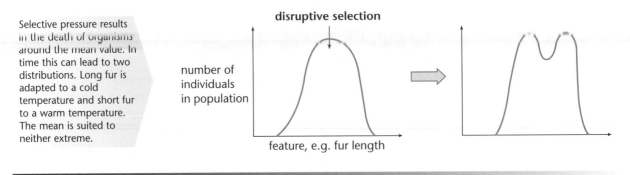

Selective pressure results in the death of organisms around the mean value. In time this can lead to two distributions. Long fur is adapted to a cold temperature and short fur to a warm temperature. The mean is suited to neither extreme.

disruptive selection

number of individuals in population

feature, e.g. fur length

Speciation

The previous example of disruptive selection showed how two extreme genotypes can be selected. Continued selection against individuals around the former mean genotype finally results in two discrete distributions. This division into two groups may be followed by, for example, advantageous mutations. There is a probability that, in time, the two groups will become incompatible and unable to breed successfully. They have become a new species. The **development of new species is called speciation**.

To enable enough genetic differences to build up between the two groups, they must be isolated to stop them breeding. This can happen in a number of ways.

Geographical isolation

This takes place when a population becomes divided as a result of a physical barrier appearing. For example, a land mass may become divided by a natural disaster like an earthquake or a rise in sea level. Geographical isolation followed by mutations can result in the formation of new species. This can be illustrated with the finches of the Galapagos islands. There are many different species in the Galapagos islands, ultimately from a common ancestral species. Clearly new species do form after many years of geographical isolation. This is **allopatric speciation**.

Reproductive isolation

This is a type of genetic isolation. Here the formation of a new species can take place in the same geographical area, e.g. mutation(s) may result in reproductive incompatibility. A new gene producing, for example, a hormone, may lead an animal to be rejected from the mainstream group, but breeding may be possible within its own group of variants. The production of a new species by this mechanism is known as **sympatric speciation**.

Sample question and model answer

The diagrams below show the transfer of a useful gene from a donor plant cell to the production of a transgenic crop plant. The numbers on the diagram show the stages in the process.

Look out for transgenic stories in the media. The principles are often the same. This will prepare you for potentially new ideas in your 'live' examinations. You could encounter the same account!

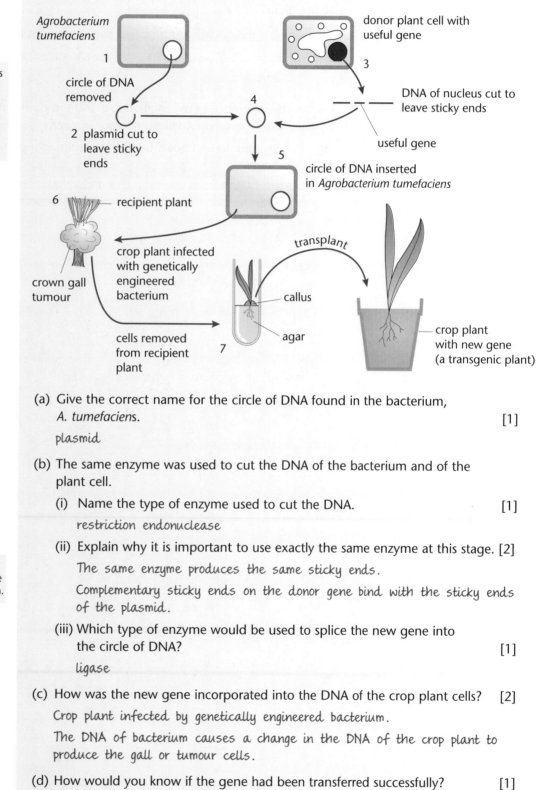

(a) Give the correct name for the circle of DNA found in the bacterium, *A. tumefaciens.* [1]

plasmid

(b) The same enzyme was used to cut the DNA of the bacterium and of the plant cell.

(i) Name the type of enzyme used to cut the DNA. [1]

restriction endonuclease

(ii) Explain why it is important to use exactly the same enzyme at this stage. [2]

The same enzyme produces the same sticky ends.

Complementary sticky ends on the donor gene bind with the sticky ends of the plasmid.

(iii) Which type of enzyme would be used to splice the new gene into the circle of DNA? [1]

ligase

This question covers key techniques in gene transfer. Be prepared for your examination.

(c) How was the new gene incorporated into the DNA of the crop plant cells? [2]

Crop plant infected by genetically engineered bacterium.

The DNA of bacterium causes a change in the DNA of the crop plant to produce the gall or tumour cells.

(d) How would you know if the gene had been transferred successfully? [1]

The feature would be expressed in the transgenic plants.

Practice examination questions

1 A new genetically modified soya bean plant has been developed. It has a new gene which prevents it from being killed by herbicide (weed killer).

(a) Describe the stages which enable a gene to be transferred from one organism to another. [5]

(b) Suggest how the genetically modified soya plants could result in higher bean yields. [3]

[Total: 8]

2 (a) Explain the difference between allopatric and sympatric speciation. In each instance use an example to illustrate your answer. [6]

(b) How is it possible to find out if two female animals are from the same species? [2]

[Total: 8]

Chapter 5

Ecology

The following topics are covered in this chapter:

- *Investigation of ecosystems*
- *Energy flow through ecosystems*
- *Global warming*

- *Colonisation and succession*
- *Nutrient cycles*

5.1 Investigation of ecosystems

After studying this section you should be able to:

- *use the capture, mark, recapture technique to assess animal populations*
- *use quadrats to map the distribution of organisms*
- *understand the factors that affect the distribution of organisms*

> **LEARNING SUMMARY**

Measurement in an ecosystem

Edexcel ▸ 4.5.10–11

The study of ecology investigates the inter-relationships between organisms in an area and their environment. The area in which organisms live is called a **habitat**. The combination of the organisms that live in a habitat and the physical aspects of the habitat is called an **ecosystem**.

Estimating populations

All the individuals of one species living together in a habitat are called a **population**. The size of plant populations can be estimated by using a quadrat placed at random. Animals, however, do not tend to stay still for long enough to be sampled using a quadrat. The population size of an animal species can be estimated by using capture–recapture.

Capture, mark, release, recapture

This is a method which is used to estimate animal populations. It is an appropriate method for motile animals such as shrews or woodlice. The ecologist must always ensure minimum disturbance of the organism if results are to be truly representative and that the population will behave as normal.

Before using the technique you must be assured that:

- there is no significant migration
- there are no significant births or deaths
- marking does not have an adverse effect, e.g. the marking paint should not allow predators to see prey more easily (or vice versa)
- organisms integrate back into the population after capture.

Remember that the method is suitable for large population size only.

The technique

- Organisms are captured, *unharmed*, using a quantitative technique.
- They are counted then discretely marked in some way, e.g. a shrew can be tagged, a woodlouse can be painted (*with non-toxic paint*).
- They are released.
- Organisms from the same population are recaptured, and another count is made, to determine the number of marked animals and the number unmarked.

The calculation

S = total number of individuals in the total population.
S_1 = number captured in sample one, marked and released, e.g. 8.
S_2 = total number captured in sample two, e.g. 10.
S_3 = total marked individuals captured in sample two, e.g. 2.

$$\frac{S}{S_1} = \frac{S_2}{S_3} \quad \text{so, } S = \frac{S_1 \times S_2}{S_3} \quad \text{population} = \frac{8 \times 10}{2} = 40 \text{ individuals}$$

Remember the equation carefully. You will **not** be supplied with it in the examination, but you will be given data.

Measuring the distribution of organisms

This can be measured using another quadrat technique called a belt transect. This method should be used when there is a **transition** across an area, e.g. across a pond or from high to low tide on the sea shore. Use belt transects where there is **change**. The belt transect is a line of quadrats. In each quadrat a measurement such as density can be made. One transect is not enough! Always do a number of transects then find an average for quadrats in a similar zone.

A bar graph would be used to show the **distribution** of plant species across the pond. Note that there would be more than just two species. The graphs show how you could illustrate the data. Clearly flag irises occupy a different niche to water lilies.

A simplified results table

Quadrat no.	flag iris	water lily
1	10	0
2	7	0
3	1	0
4	0	5
5	0	4
6	0	0
7	0	5
8	0	3
9	0	0
10	1	0
11	8	0
12	4	0

This is just one belt transect. A number would be used and an average taken for each corresponding quadrat.

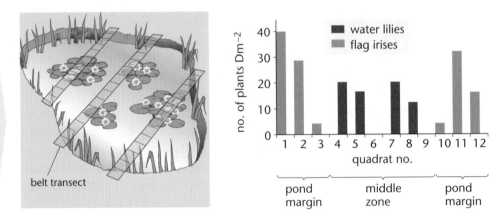

belt transect

Other uses of quadrats

Quadrats can also be used to survey animal populations. It is made easier if the organisms are **sessile** (*they do not move from place to place*), e.g. barnacles on a rock. In a pond the belt transect could be coupled with a kick sampling technique. Here rocks may be disturbed and escaping animals noted. Adding a further technique can help, such as using a catch net in the quadrat positions. The principle here is that the techniques are **quantitative**.

KEY POINT

Factors that determine population size

Graphical data can show relative numbers and distribution of organisms in a habitat. The ecologist is interested in the factors that determine the size and distribution of organisms.

These factors can be **biotic** or **abiotic**.

Abiotic factors are non-living factors. They include:

- carbon dioxide level
- oxygen level
- pH
- light intensity
- mineral ion concentration
- level of organic material.

Biotic factors are living factors.

They include:

- **Competition** This occurs when organisms are trying to get the same resources. There are two types. **Interspecific competition** takes place when **different** species share the same resources. **Intraspecific competition** takes place when the **same** species share the same resources.
- **Predation** This involves feeding relationships.

Predators and prey

There can be many examples of this type of relationship in an ecosystem. **Primary consumers** rely on the **producers**, so a flush of new vegetation may give a corresponding increase in the numbers of primary consumers. Predators which eat the primary consumers may also follow with a population increase. Each population of the ecosystem may have a **sequential effect** on other populations. Ultimately, the ecosystem is in **dynamic equilibrium** and has limits as to how many of each population can survive, i.e. its **carrying capacity**.

Note that graphs are often given in predator–prey questions. A flush of spring growth is often responsible for the increase in prey. Plant biomass may not be shown on the graph! Candidates are expected to suggest this for a mark. Also remember that as prey increase, their numbers will go down when eaten by the predator. Predator numbers rise after this.

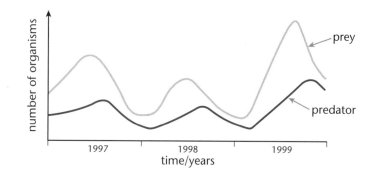

5.2 Colonisation and succession

After studying this section you should be able to:

- *understand how colonisation is followed by changes*
- *understand how colonisation and succession lead to a climax community*

LEARNING SUMMARY

How colonisation and succession take place

Edexcel   4.5.13

Any area which has never been inhabited by organisms may be available for **primary succession**. Such areas could be a garden pond filled with tap water, lava having erupted from a volcano, or even a concrete tile on a roof. The latter may become colonised by lichens. Occasionally an ecosystem may be destroyed, e.g. fire destroying a woodland. This allows **secondary succession** to begin, and signals the reintroduction of plant and animal species to the area.

Colonisation and succession also take place in water. Even an artificial garden pond would be colonised by organisms naturally. Aquatic algae would arrive on birds' feet.

- **Pioneer species (primary colonisers)** begin to exploit a 'new' habitat. Mosses may successfully grow on newly exposed heathland soil. These are the **primary colonisers** which have adaptations to this environment. Fast germination of spores and the ability to grow in waterlogged and acid conditions aid rapid colonisation. These plants may support a specific food web. In time, as organic matter drops from these herbaceous colonisers it is decomposed, nutrients are added to the soil and acidity increases. In time, the changes caused by the primary colonisers make the habitat unsuitable.
- Conditions unsuitable for primary colonisers may be ideal for other organisms. In heathland, mosses are replaced by heathers which can thrive in acid and xerophytic (desiccating) conditions. This is **succession**, where one community of organisms is replaced by another. In this example, the secondary colonisers have replaced the primary colonisers; this is known as **seral stage 1** in the succession process. Again, a different food web is supported by the secondary colonisers.

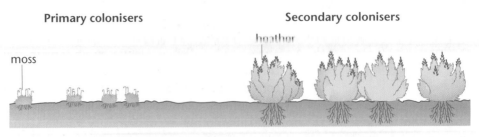

Primary colonisers   Secondary colonisers

moss   heather

- At every seral stage, there are changes in the environment. The **second seral stage** takes place as the tertiary colonisers replace the previous organisms. In heathland, the new conditions would favour shrubs such as gorse and bilberry plus associated animals.
- The shrubs are replaced in time with birch woodland, the **third seral stage**. Eventually, acidic build up leads to the destruction of the dominant plant species.
- Finally, conditions become suitable for a dominant plant species, the oak. Tree saplings quickly become established. Beneath the oak trees, grasses, ferns, holly and bluebells can grow in harmony. This final stage is **stable** and can continue for hundreds of years. This is the **climax community**. Associated animals survive and thrive alongside these plant resources. Insects such as gall wasps exploit the oak and dormice eat the wasp larvae.

Jays are birds which eat some acorns but spread others which they store and forget. The acorns germinate; the woodland spreads.

Climax community

oak woodland

In Britain, an excellent example of a climax community is Sherwood Forest where the 'Major Oak' has stood for 1000 years. Agricultural areas grow crops efficiently by **deflecting succession**. Plants and animals in their natural habitat are 'more than a match' for domesticated crops. Herbicides and pesticides are used to stop the invaders!

5.3 Energy flow through ecosystems

After studying this section you should be able to:

- *understand the roles of producers, consumers and decomposers in food chains*
- *understand the flow of energy through an ecosystem*

LEARNING SUMMARY

Food chains and energy flow

Edexcel   4.5.7–8

Before energy is available to organisms in an ecosystem, photosynthesis must take place. Sunlight energy enters the ecosystem and some is available for photosynthesis. Not all light energy reaches photosynthetic tissues. Some totally misses plants and may be absorbed or reflected by items such as water, rock or soil. Some light energy which does reach plants may be reflected by the waxy cuticle or even miss chloroplasts completely! The energy that is trapped by photosynthesis and converted into biomass is called the **gross primary productivity (GPP)**.

> Around 4% of light entering an ecosystem is actually used in photosynthesis.

KEY POINT

The green plant uses the **carbohydrate** as a first stage substance and goes on to make **proteins** and **lipids**. Plants are a rich source of nutrients, available to the herbivores which eat the plants. Some energy is not available to the herbivores because green plants **respire** (releasing energy).

The energy that is available to herbivores is called the **net primary productivity (NPP)**. It is calculated as follows:

Net primary productivity = gross primary productivity – energy lost in respiration

Energy is also lost from the food chain as **not all parts** of plants may be **consumed**, e.g. roots.

Food chains and webs

Energy is passed along a food chain. Each food chain always begins with an **autotrophic** organism (producer), then energy is passed to a primary consumer, then a secondary consumer, then a tertiary consumer and so on.

direction of energy flow →

Producer → primary consumer → secondary consumer → tertiary consumer
(herbivore)   (1st carnivore)   (2nd carnivore)

The following example shows three food chains linked to form a food web.

Note that a small bird is a secondary consumer when it eats apple codling moths but a tertiary consumer when it eats greenfly.

The producers always have more energy than the primary consumers, the primary consumers more than the secondary consumers and so on, up the food web. Energy is released by each organism as it respires. Some energy fails to reach the next organism because not all parts may be eaten.

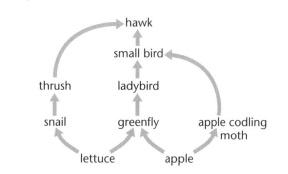

Each feeding level along a food chain can also be represented by a **trophic level**. The food chain below is taken from the food web above and illustrates trophic levels. Energy may be used by an organism in a number of different ways:

* respiration releases energy for movement or maintenance of body temperature, etc.
* production of new cells in growth and repair
* production of eggs
* released trapped in excretory products.

Examiner's tip! Note that the primary consumer is at trophic level 2. It is easy to make a mistake with this concept. Many candidates do!

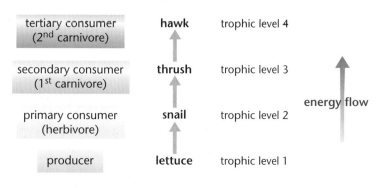

Pyramids of numbers, energy and biomass

A food chain gives limited information about feeding relationships in an area. Actual proportions of organisms in an area give more useful data. Consider a food chain from a wheat field. The pyramid of numbers sometimes does not give a suitable shape. In the example shown below, there are more aphids in the field than wheat plants. This gives the shape shown below (not a pyramid in shape!). A pyramid of biomass is more likely to be a pyramid in shape because it takes into account the size of the organism. It does not always take into account the rate of growth and so only a pyramid of energy is always the correct shape.

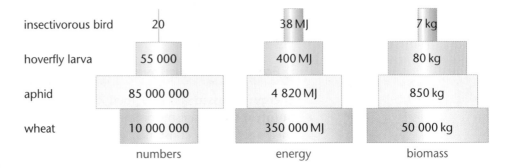

	numbers	energy	biomass
insectivorous bird	20	38 MJ	7 kg
hoverfly larva	55 000	400 MJ	80 kg
aphid	85 000 000	4 820 MJ	850 kg
wheat	10 000 000	350 000 MJ	50 000 kg

> Biomass is the mass of organisms present at each stage of the food chain. The biomass of wheat would include leaves, roots and seeds. (All parts of the plant are included in this measurement.)

The organisms in the above food chain may die rather than be consumed. When this happens, the decomposers use extra-cellular enzymes to break down any organic debris in the environment. Corpses, faeces and parts that are not consumed are all available for decay.

5.4 Nutrient cycles

After studying this section you should be able to:

- recall how carbon is recycled

L E A R N I N G
S U M M A R Y

The carbon cycle

Edexcel   4.5.9

Carbon is the key element in all organisms. The source of this carbon is atmospheric carbon dioxide which proportionally is 0.03% of the volume of the air. Most organisms cannot use carbon dioxide directly so the carbon cycle is very important.

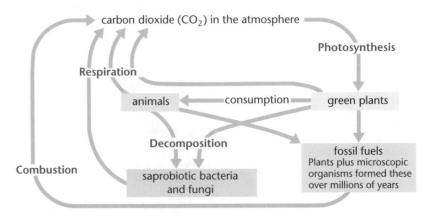

Some important points

- Producers carry out photosynthesis. This process incorporates the carbon dioxide into carbohydrates. These chemicals are used as a starting point to make lipids and proteins. Some of the carbon helps to form structures in the producers and some is released as carbon dioxide as a waste product of respiration.
- Producers are the starting point of food chains. After the plants are eaten by primary consumers carbon can be passed along to subsequent consumers. It can be incorporated into tissues, respired or excreted.
- Even after the death of a plant or animal, carbon dioxide can still be released. Saprobiotic bacteria and fungi respire using the organic chemicals in dead organisms as well as faeces and urine, etc.
- Compression of organisms millions of years ago resulted in the formation of fossil fuels. Combustion of these fuels releases carbon dioxide back into the atmosphere.
- The return of carbon dioxide to the air completes the cycle!

Other elements are also recycled. The decomposers have a major role in maintaining the availability of vital chemicals.

Examiners often give a question about what happens to the energy in the chemicals of dead organisms or organic waste. Many candidates correctly state that microorganisms rot down the materials but then go on to state that energy goes into the ground. Big mistake! Energy is released by the respiration of the decomposers to support their life.

5.5 Global warming

After studying this section you should be able to:

- *understand how the greenhouse effect could lead to global warming*
- *understand some of the evidence for the causes of global warming*

LEARNING SUMMARY

How human activities affect the environment

Edexcel 4.5.14-20

Activities carried out by the human population to supply food, power and industrial needs have considerable effects on the environment. These effects include atmospheric and water pollution, and destroying habitats and communities.

What is the greenhouse effect?

This is caused by specific gases which form a thin layer around the atmosphere. These gases include water vapour, carbon dioxide, methane, ozone, nitrogen oxides and CFCs. CFCs (CCl_2F_2, CCl_3F) have a greenhouse factor of 25 000 based on the same amount of carbon dioxide at a factor of 1.0. The **quantity** of the greenhouse factor gas needs to be considered to work out the overall greenhouse effect, e.g.

> carbon dioxide is 0.035% of the troposphere × greenhouse factor value
>   1= 0.035
> CCl_2F_2 is 4.8×10^{-8} % of the troposphere × greenhouse factor value
>   25 000 = 0.012
> Water vapour is 1% of the troposphere × greenhouse factor
>   0.1 = 0.1

KEY POINT

It is clear that water vapour has the greatest overall greenhouse effect!

Do not mix up the greenhouse effect with the 'hole in the ozone layer' – that is different! The ozone layer around the Earth absorbs some ultra-violet radiation from the sun. If a lot of ultra-violet radiation reaches the Earth's surface then many people succumb to skin cancer. CFCs cause a hole to form in the ozone layer. Not using these chemicals is the answer to this problem.

- The greenhouse gases allow short wavelength radiation from the sun to reach the Earth's surface.
- Some of the infra-red radiation fails to pass back through the greenhouse layer resulting in **global warming**.
- Polar ice caps may melt causing the sea to rise and subsequent reduction of land mass. Some aquatic populations could increase and some terrestrial populations decrease.
- Climatic changes are expected, so rainfall changes and heat increases will have significant effects.

greenhouse effect

Follow the numbers to remember sequence in correct order!

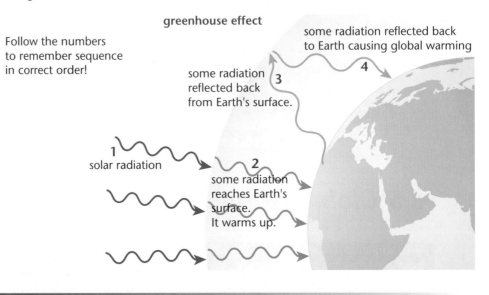

Climate change – fact or fiction?

The idea that the Earth's climate has changed over long periods in the past is a well accepted fact amongst most people but not everybody agrees on the cause.

The theory that changes in our climate could be caused by changes in the carbon dioxide levels in the atmosphere was first put forward by a Swedish scientist called Svante Arrhenius in 1889. But what is the evidence that global warming can be caused by increasing carbon dioxide levels? Scientists have used several sets of data to try to provide evidence for this theory:

- Direct records of temperatures using thermometer readings. Unfortunately, this data only goes back to about 1850.
- Pollen records in peat bogs to see which plant species were alive.
- Dendrochronology, using the size of annual growth rings in ancient trees to judge the climate.
- Ice cores, such as those in Vostok, Antarctica.

Vostok is a remote place in Antarctica where a hole has been drilled down into the ice over 3.3 kilometres deep. The ice core that has been removed contains bubbles of air that were trapped at different times over thousands of years. The gas has been analysed to give data about the levels of carbon dioxide and the temperature of the air way back into the past.

Graphs have been plotted that look like this:

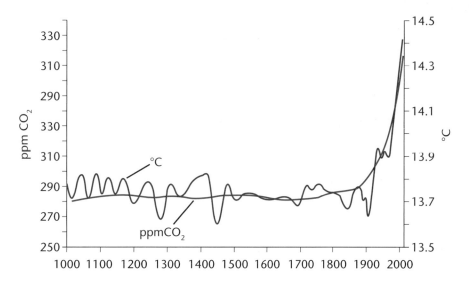

Looking at the graph, the levels of carbon dioxide and the temperature seem to follow a very similar pattern. Scientists say that there is a correlation between carbon dioxide levels and the temperature. This provides evidence for the theory but does not prove it.

Different scientists have interpreted the mass of data that is accumulating in different ways. This has caused much controversy.

Sample question and model answer

When given a passage, line numbers are often referred to. Try to understand the words in context. Do not rush in with a pre-conceived idea.

Read the passage, then answer the questions below.

line 1   Around the UK coast there are two species of barnacle, *Chthamalus stellatus* and *Balanus balanoides*. Both species are sessile, living on rocky sea shores.

The adult barnacles do not move from place to place but do reproduce
line 5   sexually. They use external fertilisation. Larvae resemble tiny crabs and are able to swim. At a later stage these larvae come to rest on a rock where they become fixed for the remainder of their lives.

The barnacles are only able to feed while submerged.

Adult *Chthamalus* are found higher on the rocks than *Balanus* in the adult
line 10 form as shown in the diagram below. Scientists have shown that the larvae of each species are found at all levels.

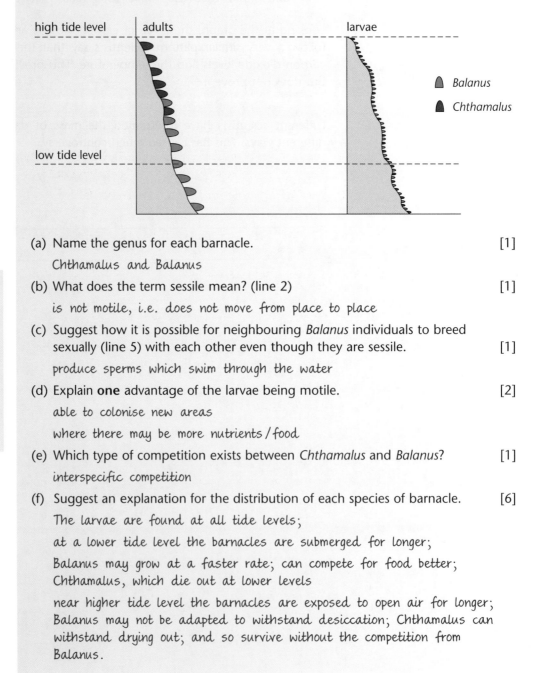

(a) Name the genus for each barnacle. [1]

    Chthamalus and Balanus

(b) What does the term sessile mean? (line 2) [1]

    is not motile, i.e. does not move from place to place

(c) Suggest how it is possible for neighbouring *Balanus* individuals to breed sexually (line 5) with each other even though they are sessile. [1]

    produce sperms which swim through the water

(d) Explain **one** advantage of the larvae being motile. [2]

    able to colonise new areas

    where there may be more nutrients / food

(e) Which type of competition exists between *Chthamalus* and *Balanus*? [1]

    interspecific competition

(f) Suggest an explanation for the distribution of each species of barnacle. [6]

    The larvae are found at all tide levels;

    at a lower tide level the barnacles are submerged for longer;

    Balanus may grow at a faster rate; can compete for food better; Chthamalus, which die out at lower levels

    near higher tide level the barnacles are exposed to open air for longer; Balanus may not be adapted to withstand desiccation; Chthamalus can withstand drying out; and so survive without the competition from Balanus.

In this question are key terms, of which you will need to recall the meaning. This is only possible with effective revision. Did you already know the key terms **genus**, **sessile** and **motile**? Knowledge of these terms would enable you to access other marks, only easy when you have key word understanding. Try writing out a glossary of terms to help your long-term memory.

1 The following two gases help cause the greenhouse effect.

	Greenhouse effect factor	Relative amount in troposphere
Water vapour	0.1	1%
CFCs	25 000	4.8×10^{-8}

(a) Work out which gas has the greatest influence on the greenhouse effect. [2]

(b) Suggest **one** reason for the greenhouse effect resulting in:

(i) an increase in the population of a species [1]

(ii) a decrease in the population of a species. [1]

[Total: 4]

2 Ecologists wished to estimate the population of a species of small mammal in a nature reserve.

- They placed humane traps throughout the reserve and made their first trapping on day one, capturing 16 shrews.
- They were tagged then released.
- After day four a second trapping was carried out, capturing 12 shrews.
- Five of these shrews were seen to be tagged.

(a) The ecologists must be satisfied of a number of factors before using the 'capture, mark, release, recapture' method. List three of these factors. [3]

(b) Use the data to estimate the shrew population.
Show your working. [2]

(c) Comment on the *level* of reliability of your answer. [1]

[Total: 6]

Chapter 6

Disease, exercise and health

The following topics are covered in this chapter:

- *Diseases*
- *Immunity*
- *Assisting the body's defences*
- *Exercise and health*

6.1 Diseases

After studying this section you should be able to:

- *describe different causes of disease*
- *explain how the body defends itself against pathogens*
- *describe how vaccines and antibiotics can be used to help defend against pathogens*
- *discuss the relationship between exercise and health*

LEARNING SUMMARY

What is a disease?

EDEXCEL   4.6.8, 4.6.10

A disease is a **disorder** of a tissue, organ or system of an organism. As a result of a disorder, **symptoms** are evident. Such symptoms could be the failure to produce a particular digestive enzyme, or a growth of cells in the wrong place. Normal bodily processes may be disrupted, e.g. efficient oxygen transport is impeded by the malarial parasite, *Plasmodium*.

Different types of disease

Genetic diseases

These can be passed from parent to offspring, e.g. haemophilia and cystic fibrosis.

Dietary related diseases

These are caused by the foods that we eat. Too much or too little food may cause disorders, e.g. obesity or anorexia nervosa. Lack of vitamin D causes the bone disease rickets, the symptoms of which are soft weak bones which bend under the body weight.

Environmentally related diseases

Some aspects of the environment disrupt bodily processes, e.g. as a result of nuclear radiation leakage, cancer may develop.

An **auto-immune disease** may be **environmentally** caused, e.g. the form of leukaemia where phagocytes destroy a person's red blood cells may be caused by radiation leakage.

Auto-immune disease

The body in some way attacks its own cells so that processes fail to function effectively.

Infectious disease by pathogens

Pathogens attack an organism and can be passed from one person to another. Many pathogens are spread by a **vector** which carries them from one organism to another without being affected itself by the disease. Pathogens include bacteria, viruses, protozoa, fungi, parasites and worms.

Viruses

Viruses are so simple in structure that they are not considered members of any kingdom. They do not breathe, feed or excrete. They can, however, replicate. The diagrams below show typical viral characteristics.

Many viruses are disease causing agents, e.g. the polio causing virus. Some viruses, e.g. the T2 bacteriophage can be helpful.

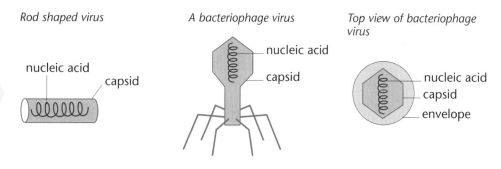

Rod shaped virus

nucleic acid

capsid

A bacteriophage virus

nucleic acid

capsid

Top view of bacteriophage virus

nucleic acid

capsid

envelope

KEY POINT

Shape is variable but they do have some common features:
- an outer coat (**capsid**) consisting of protein units (**capsomeres**)
- an internal core of **RNA** or **DNA**
- they all **reproduce** by using the DNA of a **host cell**, so in this respect they are all **parasitic**
- some viruses have an additional outer cover known as an **envelope**.

Bacteria

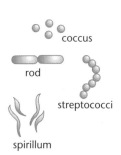

coccus

rod

streptococci

spirillum

Bacteria are members of the kingdom Prokaryotae, and include saprobiotic and parasitic species. They exist in a number of different shapes. A selection is shown here, together with a generalised structure.

Typical bacterial features

- Cell wall which is not made of cellulose.
- No true nucleus, but the DNA is in nucleiod form, a single chromosome of coiled DNA, and in circular plasmids (in some bacteria).
- If flagellae are present, there is not a 9 + 2 filament structure.
- Ribosomes are present but they are small.
- Usual reproduction by binary fission, a form of asexual reproduction.
- No membrane bound organelles.

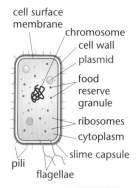

cell surface membrane

chromosome

cell wall

plasmid

food reserve granule

ribosomes

cytoplasm

slime capsule

pili

flagellae

Be prepared to answer questions about diseases not on your syllabus. The examiners will give data and other information which you will need to interpret. Use your knowledge of the principles of disease transmission, infection, symptoms and cure.

How are infectious diseases transmitted?

The pathogens which cause infectious diseases are transmitted in a range of ways.

- Direct contact – sexual intercourse enables the transmission of syphilis bacteria; a person's foot which touches a damp floor at the swimming baths can transfer the Athlete's foot fungus.
- Droplet infection – a sneeze propels tiny droplets of nasal mucus carrying viruses such as those causing influenza.
- Via a vector – if a person with typhoid bacteria in the gut handles food, the bacteria can be passed to a susceptible person.
- Via food or water – chicken meat kept in warm conditions encourages the reproduction of *Salmonella* bacteria, which are transferred to the human consumer, who has food poisoning as a result.
- Via blood transfusion – as a result of receiving infected blood, a person can contract AIDS.

Some infectious diseases have serious consequences to human life. The incidence of infectious diseases may vary according to the climate of the country, the presence of vectors, the social behaviour of people and other factors.

Disease file – tuberculosis

EDEXCEL 4.6.11

Cause of disease

Mycobacterium tuberculosis (bacterium) via droplet infection.

Transmission of microorganism

Coughs and sneezes of sufferers spread tiny droplets of moisture containing the pathogenic bacteria. People then inhale these droplets and may contract the disease.

Outline of the course of the disease and symptoms

The initial attack takes place in the lungs. The alveoli surfaces and capillaries are vulnerable and lesions occur. Some epithelial tissues begin to grow in number but these cannot carry out gaseous exchange. Inflammation occurs which stimulates painful coughing. Intense coughing takes place which can cause bleeding. There is much weight loss. Weak groups of people, like the elderly, or those underweight are more prone to the disease.

Prevention

Mycobacterium bovis causes tuberculosis in cattle. It can be passed to humans via milk. It causes an intestinal complaint in humans. It is important that cows are kept free of *M. bovis* by antibiotics.

The BCG vaccination is the injection of a weakened form of this microbe. This vaccination stimulates antibodies which are effective against both *M. tuberculosis* and *M. bovis*.

Mass screening using **X-rays** can identify 'shadows' in those people with scar tissue in the lungs.

Sputum testing identifies the presence of the bacteria in sufferers. Sufferers can be treated with antibiotics. Once cured they cannot pass on the pathogen so an epidemic may be prevented.

Skin testing is used. Antigens from dead *Mycobacteria* are injected just beneath the skin. If a person has been previously exposed to the organism then the skin swells which shows that they already have resistance, i.e. they have antibodies already. Anyone whose skin does not swell up is given the **BCG vaccination**. This contains attenuated *Mycobacterium bovis* to stimulate the production of antibodies against both *M. bovis* and *M. tuberculosis*.

Cure

Use of antibiotics such as streptomycin.

Disease file – AIDS (Acquired Immune Deficiency Syndrome)

EDEXCEL 4.6.11

Cause of disease

This is caused by HIV (human immune deficiency virus). It is a retrovirus, which is able to make DNA with the help of its own core of RNA.

Transmission of microorganism

This takes place by the exchange of body fluids, transfusion of contaminated blood, or via syringe needle 'sharing' in drug practices.

Outline of the course of the disease and symptoms

Scientists are constantly trying to find a **cure**. None has been found yet.

Destruction of T-lymphocyte cells

The HIV protein coat attaches to protein in the plasma membrane of a T-lymphocyte. The virus protein coat fuses with the cell membrane releasing RNA and reverse transcriptase into the cell. This enzyme causes the cell to produce DNA from the viral RNA. This DNA enters the nucleus of the T-lymphocyte and is incorporated into the host cell chromosomes. The gene representing the HIV virus is permanently in the nucleus from now on and can be dormant for years. It may become activated by an infection. Viral protein and viral RNA are made as a result of the infection.

Many RNA viral cores now leave the cell and protein coats are assembled from degenerating plasma membranes. Other T-lymphocytes are attacked. Cells of the lymph nodes and spleen are also destroyed. Viruses appear in the blood, tears, saliva, semen and vaginal fluids. The immune system becomes so weak that many diseases can now successfully invade the weakened body.

Prevention

Screening of blood before transfusions. Use of condoms and remaining with one partner. No use of contaminated needles.

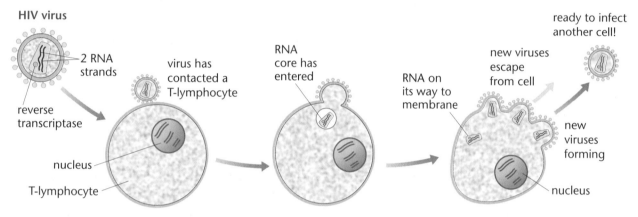

6.2 Immunity

After studying this section you should be able to:

- *describe the mechanisms used by the body to try and prevent pathogen entry*
- *describe and explain the action of the body's immune system*

LEARNING SUMMARY

Survival against the attack of pathogens

EDEXCEL 4.6.10

Many pathogenic organisms attack people. They are not all successful in causing disease. We have immunity to a disease when we are able to resist infection. The body has a range of ways to prevent the disease-causing organism from becoming established.

- A tough protein called **keratin** helps skin cells to be a formidable **barrier** to prevent pathogens entering the body.
- An enzyme, **lysozyme**, destroys some microorganisms and can be found in sebum, tears and saliva.
- **Hydrochloric acid** in the stomach kills some microorganisms.
- The bronchial tubes of the lungs are lined with **cilia**. Microorganisms which enter the respiratory system are often trapped in mucus which is then moved to the oesophagus. From here they move to the stomach where many are destroyed by hydrochloric acid or digested.
- **Blood clotting** in response to external damage prevents entry of microorganisms from the external environment.

The methods the body uses to prevent microorganisms entering the bloodstream are sometimes unsuccessful. When the microorganisms invade and then breed in high numbers, we develop the symptoms. **White blood cells** enable us to destroy invading microorganisms. They may destroy the microorganisms quickly before they have any chance of becoming established, so the person would not develop any symptoms. Sometimes there are so many microorganisms attacking that the white blood cells cannot destroy all of them. Once the pathogens are established the symptoms of a disease follow, but for most diseases, after some time, the white blood cells eventually overcome the disease-causing organisms.

The roles of the white blood cells (leucocytes)

EDEXCEL   4.6.12

There are a number of different types of **leucocytes**. Most are produced from **stem cells** in the **bone marrow**. Different stem cells follow alternative maturation procedures to produce a range of leucocytes. Leucocytes have the ability to recognise self chemicals and non-self chemicals. Only where non-self chemicals are recognised will a leucocyte respond. Proteins and polysaccharides are typical of the complex molecules which can trigger an immune response.

Phagocytes

> Inflammation increases the blood flow to infected areas and makes it easier for phagocytes to reach the pathogen.

Phagocytes can move to a site of infection through capillaries, tissue fluid and lymph as well as being found in the plasma. They move towards pathogens which they destroy by the process of phagocytosis. This is often called engulfment and involves the surrounding of a pathogen by pseudopodia to form a food vacuole. Hydrolytic enzymes from lysosomes complete the destruction of the pathogen.

> **Neutrophils** are one type of phagocyte. Proteins in plasma called **opsonins** attach to a pathogen. These opsonins enable the phagocyte to engulf the pathogen.
>
> **Macrophages** are another type of phagocyte which work alongside T-lymphocytes.

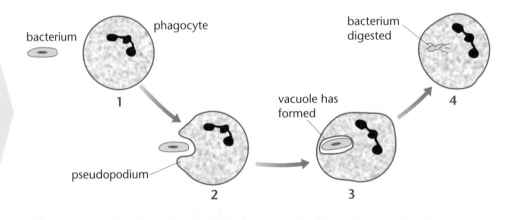

What is an antigen?

EDEXCEL   4.6.13–14

As an individual grows and develops, complex substances such as proteins and polysaccharides are used to form cellular structures. Leucocytes identify these substances in the body as 'self' substances. They are ignored as the leucocytes encounter them daily. 'Non-self substances', e.g. foreign proteins which enter the body, are immediately identified as 'non-self'. These are known as **antigens** and trigger an immune response.

Lymphocytes

> White blood cells (leucocytes) constantly check out proteins around the body. Foreign protein is identified and attack is stimulated.

In addition to phagocytes, there are other leucocytes called lymphocytes. There are two types of lymphocyte, **B-lymphocytes** and **T-lymphocytes**.

B-lymphocytes begin development and mature in the bone marrow. They produce antibodies, known as the **humoral response**.

T-lymphocytes work alongside phagocytes known as macrophages; this is known as the **cell-mediated response**. A macrophage engulfs an antigen. This antigen remains on the surface of the macrophage. T-lymphocytes respond to the antigen, dividing by mitosis to form a range of different types of T-lymphocyte cells.

- **Killer T-lymphocytes** adhere to the pathogen, secrete a toxin and destroy it.
- **Helper T-lymphocytes** stimulate the production of antibodies.
- **Suppressor T-lymphocytes** are inhibitors of the T-lymphocytes and plasma cells. Just weeks after the initial infection, they shut down the immune response when it is no longer needed.
- **Memory T-lymphocytes** respond to an antigen previously experienced. They are able to destroy the same pathogen before symptoms appear.

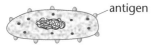

T-lymphocytes also produce a chemical called interferon which interferes with viral replication.

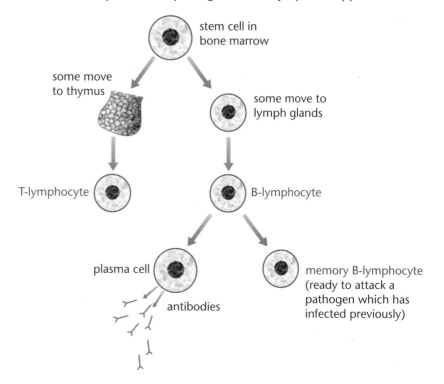

How do antibodies destroy pathogens?

The diagram on the left shows antibodies binding to antigens. The descriptions below show what can happen immediately after the binding takes place.

There are three main ways in which antibodies destroy pathogens.

- **Precipitation**, by linking many antigens together. This enables the phagocytes to engulf them.
- **Lysis**, where the cell membrane breaks open, killing the cell.
- **Neutralisation** of a chemical released by the pathogen, so that the chemical is no longer toxic.

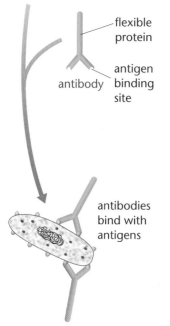

6.3 Assisting the body's defences

After studying this section you should be able to:

- *describe the origin of different types of immunity*
- *explain how immunity can be stimulated by vaccinations*
- *describe the actions of antibiotics*
- *explain how antibiotic resistance has developed*

LEARNING SUMMARY

Stimulating the immune system

EDEXCEL   4.6.15

Newborn babies are naturally protected against many diseases such as measles and poliomyelitis. This is because they have received antibodies from their mother in two ways:

- across the placenta
- via breast feeding.

This is one major advantage of breast feeding over bottle feeding of babies.

This type of immunity is called **passive natural immunity**. It is possible to give people injections of antibodies that have been made by another person or animal. This type of immunity is called **passive artificial** immunity.

> **KEY POINT**
>
> Passive immunity does not last very long. This is because the antibodies do not persist in the blood for very long.

Active immunity

Once a person has recovered from certain diseases, e.g. measles, they rarely contract that disease again. This is because the memory cells formed from B-lymphocytes can survive in the blood for many years. If the antigen reappears, they can rapidly produce a clone of antibody producing cells. This **secondary response** is rapid and larger than the **primary response** and the antigen is rapidly destroyed.

This type of immunity is called **active natural immunity**.

Vaccines consist of dead or weakened (**attenuated**) forms of pathogens. They stimulate the production of memory cells so that the person develops active artificial **immunity**.

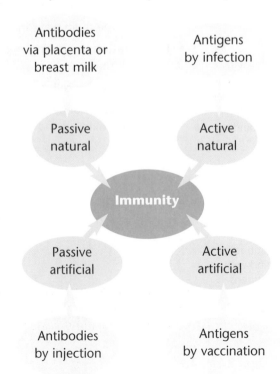

Antibodies
via placenta or
breast milk

Antigens
by infection

Passive
natural

Active
natural

Immunity

Passive
artificial

Active
artificial

Antibodies
by injection

Antigens
by vaccination

Unfortunately some pathogens show **antigenic variation**. This change in antigens renders vaccinations ineffective after a while and allows diseases such as influenza to strike many times in slightly different forms.

Scientists are now using antibody producing cells that have been fused with tumour cells. This produces cells that make large quantities of one type of antibody. They are called monoclonal antibodies. Because of their ability to target particular antigens, they might be useful in delivering drugs straight to target cells.

Antibiotics

EDEXCEL   4.6.17, 4.6.19

Antibiotics are substances produced by microorganisms which kill or inhibit further growth of other microorganisms. There are a number of different antibiotics, including penicillin.

- Discovery of penicillin was by Sir Alexander Fleming in 1928.
- He grew the bacterium *Staphylococcus* on agar.
- *Penicillium notatum* spores had reached the agar and had begun to grow.
- Next to the fungal mycelium a gap remained where no bacteria would grow.
- He found that a substance had been secreted by the fungus which he named penicillin.
- He followed this up with further research to show that the penicillin killed a number of pathogenic organisms.

> Antibiotics can be either bactericidal or bacteriostatic in action.
>
> **Bactericidal** is the term used when they kill the microbe which they attack, e.g. penicillin is bactericidal.
>
> **Bacteriostatic** is the term used when antibiotics inhibit further microbial growth.

KEY POINT

Antibiotic resistance

The term **antibiotic resistance** is used to describe the situation when a strain of bacteria has developed the ability to survive exposure to an antibiotic. This ability develops by **natural selection**:

A bacterium is produced with a mutation that provides resistance to a specific antibiotic.

Examiners like to ask synoptic questions about antibiotic resistance as it covers both this topic and also natural selection.

This antibiotic is used and it kills all the bacteria except the resistant one.

The resistant bacterium divides by binary fission and produces a resistant strain.

These strains are often referred to as **superbugs**.

Recently, strains of bacteria have been appearing that are resistant to many different types of antibiotic. These strains are becoming difficult to treat and the battle is on to find new antibiotics that are effective.

In order to prevent the spread of antibiotic resistant strains, a number of precautions should be taken:

- Antibiotics should only be available on prescription and should not be prescribed for minor illnesses or those caused by viruses.
- Patients must finish the dose. They may feel better because most of the bacteria have been killed but some may still be alive.
- The use of antibiotics should be rotated so that strains resistant to a particular antibiotic do not develop.
- The cleaning and disinfection of public places, particularly hospitals, should be improved.

Antibiotic resistance, TB and HIV

Until the development of antibiotics there were no medicines to cure TB. When antibiotics were first used many thought that TB would disappear. Now, strains of TB bacteria that are resistant to a single antibiotic have been found in almost every country.

More worryingly, a particularly dangerous form of drug-resistant TB called multidrug-resistant TB (MRR-TB) has appeared. This is caused by TB bacteria that are resistant to isoniazid and rifampicin, the two most powerful anti-TB antibiotics. Drug-resistant TB is generally treatable but it requires lengthy treatment with a large number of different drugs.

Recently, extensively drug-resistant (XDR) TB has appeared, in which the bacteria are resistant to many of the other antibiotics that are used. People who are infected with HIV are more likely to contract TB. If this is XDR then it would be particularly difficult to treat.

6.4 Exercise and health

After studying this section you should be able to:

- *explain some of the benefits and dangers of exercise*
- *describe some medical techniques that can be used to treat exercise damage*
- *discuss some of the ethical issues involving science and sport*

LEARNING SUMMARY

Exercise

EDEXCEL 5.7.18

What is exercise?

Physical activity and **exercise** do not mean the same thing. Physical activity is body movement produced by the contraction of skeletal muscle that increases energy use. The term exercise is used to describe planned, structured, repetitive physical activity that aims to improve or maintain physical fitness.

Advantages and disadvantages of exercise

Not everybody benefits equally from exercise and it is important to work out the correct level of exercise for each person. If exercise is performed correctly, it can have many **advantages**:

- Exercise is important for maintaining **physical fitness**. This can help maintain a healthy weight, healthy bone density, muscle strength and joint mobility.
- Exercise also reduces levels of **cortisol**. Cortisol is a stress hormone that causes many health problems, both physical and mental.

• Frequent exercise has been shown to help prevent serious and life-threatening conditions such as **high blood pressure, coronary heart disease, type 2 diabetes, insomnia** and **depression**.

Most of the problems caused by exercise result from either people choosing the incorrect level of exercise for their body when they start, or exercising for too long. The **problems** can include:

• **Musculoskeletal damage** – this includes traumatic (acute) injuries or overuse (chronic) injuries. Acute injuries can include torn ligaments or muscle strains. Chronic injuries develop slowly, as a result of repetitive use, causing damage to the muscles, tendons, bones or joints because there is not enough time for proper healing. Examples include tendonitis, tennis elbow and shin splints.

• **Immune suppression** – whilst moderate exercise is thought to boost the immune system, there is some evidence that over-training can lead to the immune system being damaged. This may result in the person experiencing frequent and reoccurring infections, for example, repeated colds, flu and herpes.

Young cricketers are only allowed to bowl a certain number of overs to try to prevent stress injuries.

Repairing the damage

EDEXCEL 5.7.19

People now have a longer life expectancy and more leisure time for sport. This has meant that there is an increased demand for medical treatment for musculoskeletal injuries.

Advances in medical technology have meant that recovery from these injuries can be faster and more complete. Examples of these advances include:

Using keyhole surgery to repair damaged cruciate ligaments in the knee

The ligament may be torn due to a fall or a sports injury. This makes the knee unstable. The aim of the operation is to reconstruct the torn ligament with one or more tendons taken from elsewhere in the person's body. A viewing scope may be used so that the surgeon only has to make a small incision.

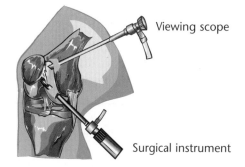

Viewing scope

Surgical instrument

Knee joint replacement using prosthetics

A prosthesis is a device designed to replace a missing or damaged part of the body. In the knee joint, damage can occur to the cartilage due to disease or ageing. Surgery replaces the damaged tissues with a metal prosthetic.

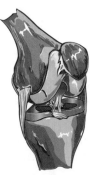

Before   After

The ethics of science in sport

CDCXCLL 5.7.20

The considerable advances in science have resulted in some difficult ethical decisions having to be made, in terms of what is and what is not allowed in sport. Two areas of debate are:

- Performance-enhancing drugs
- Prosthetics

Performance-enhancing drugs

There are a number of categories of drugs that could be used by competitors:

Type of drug	Effect on competitor	Possible side-effects
Anabolic steroids	Help athletes to train harder and increase muscle growth	Liver damage and reproductive problems
Diuretics	Used to remove fluid from the body to decrease weight, e.g. in boxing	Dehydration
Narcotic analgesics (painkillers)	Cover / mask the pain caused by injury	Make the damage worse or cause addiction
Peptides	EPO (Erythropoietin) increases red blood cell count	EPO – Blood is thicker, which can lead to a stroke
	HGH (Human growth hormone) builds muscle and bone	HGH – Abnormal growth
Stimulants	Make athletes more alert and mask fatigue	Can cause heart failure or addiction

Some people believe that these drugs should be banned to try to make competition fair. Others think they should be banned in order to protect the competitors. A counter viewpoint is that sport is not an equal competition anyway, as some competitors have better training facilities, etc., so banning drugs would still not create a 'level playing field'.

Prosthetics

Recently there has been some debate concerning the use of certain prosthetics in sport. The South African athlete Oscar Pistorius uses prosthetic devices, called blades, to run. He has competed very successfully in events against other disabled athletes but wants to be allowed to compete in all events. The question is, do his prosthetics simply act as replacements for his missing limbs or do they give him an unfair advantage?

Sample question and model answers

The graph below shows the relative numbers of antibodies in a person's blood after the vaccination of attenuated viruses. Vaccinations were given on day 1 then 200 days later.

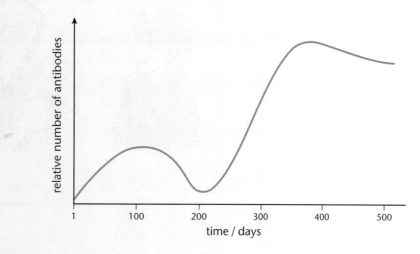

In examinations you are regularly given graphs. Make sure that you can link the idea being tested. This should help you recall all of the important concepts needed. All you need to do after this is **apply** your knowledge to the given data.

(a) Why is it important that viruses used in vaccinations are attenuated? [1]

   If they were active then the person would contract the disease.

(b) Suggest **two** advantages of giving the second vaccination. [2]

   A greater number of antibodies were produced.
   The antibodies remain for much longer after the second vaccination.

(c) Which cells produced the antibodies during the primary response? [1]

   B-lymphocytes.

(d) Why was there no delay in the secondary response to vaccination? [3]

   Because the first vaccination had already been given, memory B-lymphocytes had been produced which respond to the viruses more quickly.

(e) Describe how a virus stimulates the production of antibodies? [3]

   Antigen in the protein 'coat' or capsomere stimulate the B-lymphocytes.

If you gave B-lymphocytes as a response it would be wrong! B-lymphocytes secrete antibodies.

(f) Apart from producing antibodies, outline FOUR different ways that the body uses to destroy microorganisms. [4]

   Phagocytes by engulfment; T-lymphocytes attach to microorganisms and destroy them; hydrochloric acid in the stomach; lysozyme in tears.

Practice examination questions

1 The illustration shows how the bacterium which causes TB can be transmitted from one person to another.

(a) Name the method of disease transmission shown in the illustration. [1]

(b) Sometimes the bacteria infect people but they do not develop symptoms.

    (i) What term is given to this group of people? [1]

    (ii) Explain why these people may be a greater danger to a community than those who actually suffer from the disease. [2]

(c) (i) What can be given to a person infected with TB to help destroy the bacteria? [1]

    (ii) Explain the role of each of the following in destroying TB bacteria.

    Phagocyte
    B-lymphocyte
    T-lymphocyte [6]

2 The diagram shows the response of B-lymphocytes to a specific antigen.

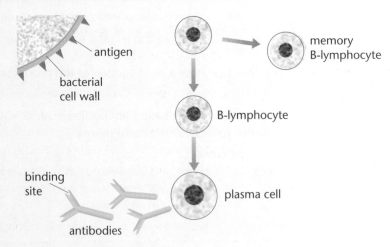

(a) (i) A plasma cell is bigger than a B-lymphocyte.
Suggest an advantage of this. [1]

    (ii) Describe the precise role of antibodies in the immune response. [3]

    (iii) What is the advantage of memory B-lymphocytes? [2]

(b) What is an auto-immune disease? Give an example. [2]

(c) A person contracts the virus which causes the common cold.
Suggest why their lymphocytes may fail to destroy the pathogen. [1]

Chapter 7

Synoptic assessment

What is synoptic assessment?

You must know the answer to this question if you are to be fully prepared for your A2 examinations!

Synoptic assessment:

- involves the drawing together of knowledge, understanding and skills learned in different parts of the AS/A2 Biology courses
- requires that candidates apply their knowledge of a number of areas of the course to a variety of contexts
- is tested at the end of the A2 course both by assessment of investigative/practical skills and by examinations
- is valued at 20% of the marks of the course total.

Each Examination Group identifies which parts of its specification will be tested in the end-assessed synoptic questions. In the Edexcel examination, synoptic questions will appear in both Unit 4 and Unit 5 examinations.

Practical investigations

You will need to apply knowledge and understanding of the concepts and principles, learned throughout the course, in the **planning**, **execution**, **analysis** and **evaluation** of each investigation.

How can I prepare for the synoptic questions?

- **Check out the modules** which will be examined for your specification's synoptic questions.

- Expect **new contexts** which draw together lots of different ideas.

- Get ready to **apply** your knowledge to a new situation; contexts change but the **principles remain the same**.

- In modular courses there is sometimes a tendency for candidates to learn for a module, achieve success, then forget the concepts. Do not allow this to happen! **Transfer concepts** from one lesson to another and from one module to another. Make those connections!

- Improve your powers of analysis – **take a range of different factors into consideration** when making conclusions; synoptic questions often involve both graphical data and comprehension passages.

- Less able candidates make limited conclusions; high ability candidates are able to consider several factors at the same time, then make a **number of sound conclusions** (not guesses!).

- You need to do **regular revision** throughout the course; this keeps the concepts 'hot' in your memory, 'simmering and distilling', ready to be **retrieved** and **applied** in the synoptic contexts.

- The bullet point style of this book will help a lot; back this up by summarising points yourself as you make notes.

Why are synoptic skills examined?

Once studying at a higher level or in employment, having a narrow view, or superficial knowledge of a problem, limits your ability to contribute. Having discrete knowledge is not sufficient. You need to have confidence in applying your skills and knowledge.

Synoptic favourites

The final modules, specified by Edexcel for synoptic assessment, include targeted synoptic questions. Concepts and principles from earlier modules will be tested together with those of the final modules. You can easily identify these questions, as they will be longer and span wide-ranging ideas.

> **Can we predict what may be regularly examined in synoptic questions?**
>
> 'Yes we can!' Below are the top five concepts. Look out for common processes which permeate through the other modules. An earlier module will include centrally important concepts which are important to your understanding of the rest.
>
> **KEY POINT**

Check out the synoptic charts

1 Energy release

Both aerobic and anaerobic respiration release energy for many cell processes. Any process which harnesses this energy makes a link.

Examples

Synoptic links

Try this yourself! Think logically. Write down an important biological term such as 'cell division'. Link related words to it in a 'flow diagram' or 'mind map'. The links will become evident and could form the framework of a synoptic question.

- Reabsorption of glucose involves active transport in the proximal tubule of a kidney nephron. If you are given a diagram of tubule cells which show both mitochondria and cell surface membrane with transporter proteins, then this is a cue that active transport will probably be required in your answer.

- Contraction of striated (skeletal) muscle requires energy input. This is another link with energy release by mitochondria and could be integrated into a synoptic question.

- The role of the molecule ATP as an energy carrier and its use in the liberation of energy in a range of cellular activity may be regularly linked into synoptic questions. The liberation of energy by ATP hydrolysis to fund the sodium pump action in the axon of a neurone.

- The maintenance of proton gradients by proton pumps is driven by electron energy. Any process involving a proton pump can be integrated into a synoptic question.

Energy: input and output

This has to be a favourite for many synoptic questions. Energy is involved in so many processes that the frequency of examination will be high.

2 Energy capture

Photosynthesis is responsible for availability of most organic substances entering ecosystems. It is not surprising that examiners may explore knowledge of this process and your ability to apply it to ecological scenarios.

Examples

- Given the data of the interacting species in an ecosystem you may be given a short question about the mechanism of photosynthesis then have to follow the energy transfer routes through food webs.

- Often both photosynthesis and respiration are examined in a synoptic type question. There are similarities in both the thylakoid membranes in chloroplasts and cristae of mitochondria.

- Many graphs in ecologically based questions show the increase in herbivore numbers, followed by a corresponding carnivore increase. Missed off the graph, your knowledge of a photosynthetic flush which stimulates herbivore numbers may be expected.

3 The structure and role of DNA

It is important to know the structure of DNA because it is fundamentally important to the maintenance of life processes and the transfer of characteristics from one generation of a species to the next. DNA links into many environmentally and evolutionally based questions.

- The ultimate source of variation is the mutation of DNA. Questions may involve the mechanism of a mutation in terms of DNA change and be followed by natural selection. This can lead to extinction or the formation of a new species. Clearly there are many potential synoptic variations.

- DNA molecules carry the genetic code by which proteins are produced in cells. This links into the production of important proteins. The structure of a protein into primary, secondary, tertiary and quaternary structure may be tested. All enzymes are proteins, so a range of enzymically based question components can be expected in synoptic questions.

- The human genome project is a high-profile project. The uses of this human gene 'atlas' will lead to many developments in the coming years. The reporting of developments, radiating from the human genome project, could be the basis of many comprehension type questions, spanning diverse areas of Biology. Save newspaper cuttings, search the internet and watch documentaries. Note links with genetic diseases, ethics, drugs, etc.

4 Structure and function of the cell surface membrane

There are a range of different mechanisms by which substances can cross the cell surface membrane. These include diffusion, facilitated diffusion, osmosis, active transport, exocytosis and pinocytosis. Additionally glycoproteins have a cell recognition function and some proteins are enzymic in function. Knowledge of these concepts and processes can be tested in cross-module questions.

- In an ecologically based question the increasing salinity of a rock pool in sunny conditions could be linked to water potential changes in an aquatic plant or animal. Inter-relationships of organisms within a related food web could follow, identifying such a question as synoptic.

- In cystic fibrosis a transmembrane regulator protein is defective. A mutant gene responsible for the condition codes for a protein with a missing amino acid. This can link to the correct functioning of the protein, the mechanism of the mutation and the functioning of the DNA.

5 Transport mechanisms

This theme may unify the following into a synoptic question, transport across membranes, transport mechanisms in animal and plant organs. Additionally, they may be linked to homeostatic processes.

- The route of a substance from production in a cell, through a vessel to the consequences of a tissue which receives the substance, could expand into a synoptic question. Homeostasis and negative feedback could well be linked into these ideas.

Sample questions and model answers

A short question which cuts across the course. It refers back to AS. Do not forget those modules! See the *Letts AS Guide* for additional advice and those concepts not found in this volume.

Question 1 (a short structured question)

The kangaroo rat (*Dipodomys deserti*) is a small mammal that lives in the Californian desert. It has specialised kidneys so that it can produce a very concentrated urine.

(a) Name the genus that contains the kangaroo rat. [1]

Dipodomys

(b) What is the biological naming system called that gives the kangaroo rat its scientific name? [1]

binomial system

(c) Kangaroo rats have long loops of Henle. In which part of the kidney would you expect to find loops of Henle? [1]

medulla

(d) What is the name of the hormone that controls the concentration of the urine in mammals? [1]

ADH

(e) Which gland releases this hormone into the blood? [1]

pituitary gland

(f) The desert community that contains the kangaroo rat is the final product of succession in California. What is the name of the final, stable community that is produced by succession? [1]

a climax community

[Total: 6]

Question 2 (a longer, more open-ended question)

Plants and animals both need to exchange gases with the environment. Describe how animals and plants are adapted for efficient gaseous exchange. [10]

(Quality of written communication assessed in this answer.)

- examples of respiratory surfaces in animals:
  gills/lungs;
  tracheoles in insects;
  surface of protoctists;
  stomata in plants
- large surface area:
  way(s) in which this is achieved e.g.
  many alveoli;
  surface area/volume ratio in protoctists;
  many gill filaments;
  large surface area of leaves;
  many mesophyll cells
- maintenance of diffusion gradients
  way(s) in which this is achieved
  rich blood supply;
  ventilation mechanisms;
  sub-stomatal airspaces;
  spongy mesophyll air spaces;
  use of carbon dioxide in mesophyll cells

- small diffusion pathway
  barriers one cell thick;
  specialised cells, e.g. squamous epithelium;
  thin cell walls of palisade cells

Note, there is one mark available for legible text with accurate spelling, punctuation and grammar.

[Total: 10]

Question 3 (a longer question of higher mark tariff)

Prepare yourself for this type of synoptic question. It cuts across a large part of the specification. Make the links with different ideas. This fact is very important; concepts from AS are needed.

Different concentrations of maltase were injected in the small intestine of a mouse. The amount of glucose appearing in the blood and the small intestine after 15 minutes were measured. The results are shown in the graph.

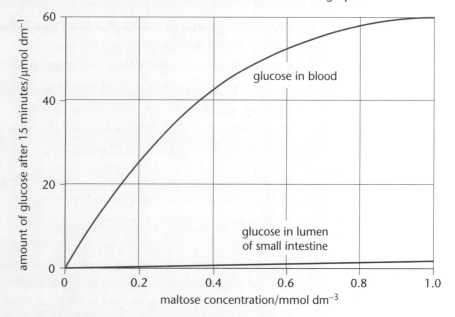

(a) (i) Describe the structure of a maltose molecule [2]

   Two molecules of (alpha) glucose;

   joined together by a glycosidic bond.

   (ii) Maltose is converted into glucose by a hydrolysis reaction. What is a hydrolysis reaction? [1]

   A reaction that breaks down a substance by the addition of water.

(b) Describe the effect of different maltose concentrations on the amount of glucose found in the lumen of the small intestine compared to the effect on the amount found in the blood. [2]

the maltose concentration has much more effect on the amount of food in the blood; the amount of glucose found in the blood is starting to level off but the amount in the lumen is increasing steadily.

Even if you only cover one of these points, you can pick up a second mark by correctly using figures from the graph in your answer.

Sample questions and model answers (continued)

(c) The enzyme maltose is found on the cell surface membrane of the epithelial cells of the small intestine.

   (i) How does the data on the graph indicate that the enzyme is not released into the lumen?   [1]

     Very little/no increase in the amount of glucose in the lumen.

This is a harder stretch and challenge question.

   (ii) Explain why having the enzyme fixed to the cell surface will increase the rate of glucose absorption.   [2]

     Higher concentration of glucose produced close to intestinal lining;

     will increase the concentration gradient between intestine and blood.

                        [Total: 8]

Question 4

A cow is described as a ruminant. Ruminants are herbivores that have a chamber in their intestines called a rumen.

(a) (i) The rumen of cows contains microorganisms.

     Explain the importance of these microorganisms to the cow.   [3]

     They digest cellulose in the cow's food;

     the cow cannot produce the enzyme to digest cellulose;

     produce fatty acids that the cow can use.

   (ii) After the food has been in the rumen for some time it is regurgitated back to the mouth for a second chewing.

     Suggest why this is important.   [1]

     Increase the surface area for digestion.

   (iii) The microorganisms in the rumen produce two waste products, methane and ammonia. The ammonia is converted into urea by the cow's liver.

     Why is this conversion important for the cow?   [1]

     Ammonia is more toxic than urea.

(b) The table shows the amount of methane produced by different domesticated animals

Animal type	Methane production per animal in kg per animal per year	Total methane production in tonnes per year
buffaloes	50	6.2
camels	58	1.0
goats	5	2.4
sheep	6	3.4

   (i) Which of the animals in the table are ruminants?
Explain how you can tell this.   [2]

     Buffaloes and camels;

     they produce much more methane per animal.

This is a typical synoptic question as it links two different topics, digestion in herbivores and the greenhouse effect!

(ii) Which type of animal in the table is domesticated in the highest numbers? Explain how you worked out your answer. [2]

Sheep; dividing the total methane production by the production per animal gives the highest number.

(iii) Methane is a potent greenhouse gas. What is a greenhouse gas? [2]

A gas that prevents the escape of infra red radiation from the atmosphere; therefore causes the atmosphere to warm.

(iv) It has recently been discovered that methane is released when arctic ice melts. Explain why people are concerned by this discovery. [2]

The release of methane would increase global warming;

which in turn would result in the release of even more methane.

[Total: 13]

Practice examination answers

Chapter 1 Energy for life

1

(a) in cytoplasm [1]

(b) pyruvate [1]

(c) 2 ATPs begin the process;
2ATPs are produced from each of the two GP molecules, so −2 + 4 = +2 ATPs net [1]

(d) animal; animal cells produce lactate [1]

(e) oxygen or aerobic [1]

[Total: 5]

2

(a) mitochondrion [1]

(b) NADH [1]

(c) cytochrome [1]

(d) ATP [1]

[Total: 4]

Chapter 2 Response to stimuli

1

(a) (i) IAA (at these lower) concentrations is *proportional* to the angle of curvature of the stem. [1]

  (ii) IAA (at these higher) concentrations is *inversely proportional* to the angle of curvature. [1]

(b) *More* IAA causes the cells at side of stem in contact with agar block to elongate more than other side.

  So this side grows more strongly bending stem towards the weaker side. [2]

(c) Growth is only stimulated up to a certain high IAA concentration, after this curvature would be inhibited. [2]

[Total: 6]

2

(a) A = actin
  B = myosin [2]

(b) action potential reaches sarcomere [1]

(c) both filaments slide alongside each other;
they form cross bridges;
during contraction the filaments slide together to form a shorter sarcomere [2]

[Total: 5]

3

(i)   resting potential achieved; [2]
  Na^+ / K^+ pump is on

(ii) Na^+ / K^+ pump is off;
  so Na^+ ions enter axon [2]

(iii) maximum depolarisation achieved; K^+ ions leave [2]

(iv) Na^+ ions leave due to Na^+/ K^+ pump being back on;
  this is during the refractory period;

(v) at end of this resting potential re-established;
  axon membrane re-polarised [4]

[Total: 10]

Chapter 3 Homeostasis

1 (a)

	Nervous system	Endocrine system
Usually have longer lasting effects		✓
Have cells which secrete transmitter molecules	✓	
Cells communicate by substances in the blood plasma		✓
Use chemicals which bind to receptor sites in cell surface proteins	✓	✓
Involve the use of Na^+ and K^+ pumps	✓	

[2]

(b) homeostasis [1]

[Total: 3]

2

(a) 86 beats per minute; 120cm³ [2]

(b) easier to measure / standardise the work done [1]

(c) increase to a peak at 130cm³ / at 140W;
increase is double resting rate;
then falls slightly at higher work rates; [3]

(d) five from:
increased respiration rate in muscles;
carbon dioxide levels rise in the blood;
detected by chemoreceptors;
in the aortic body / carotid bodies;
cardiac centre in medulla is stimulated;
increased frequency of impulses sent via accelerator nerve;
to the SA node; [5]

Chapter 4 Genetics, gene technology and selection

1

(a) Identify the specific section of DNA which contains the gene; this can be done using reverse transcriptase; insert DNA into a vector/insert into *Agrobacterium tumefaciens*; this bacterium/this vector then passes the DNA into the recipient cell. [5]

(b) herbicide kills weeds; which reduces competition; for light or water or minerals; soya plants unharmed [3]

[Total: 8]

2

(a) **Allopatric speciation** takes place after geographical isolation;
 • the rising of sea level splits a population of animals; formerly connected by land creating two islands;
 • mutations take place so that two groups result in different species.

Sympatric speciation takes place through genetic variation;
 • in the same geographical area;
 • mutation may result in reproductive incompatibility;
 • perhaps a structure in birds may lead to a different song being produced by the new variant;
 • this may lead to the new variant being rejected from the mainstream group;
 • breeding may be possible within its own group of variants. [6]

(b) Mate them both with a similar male, to give them a chance to produce fertile offspring.
 • If they both produce offspring, take a male and female from the offspring, mate them,
 • if they produce fertile offspring then original females **are** from the same species. [2]

[Total: 8]

Chapter 5 Ecology

1

(a) greenhouse effect factor x amount
 water vapour $0.1 \times 1 = 0.1$
 CFCs $25\,000 \times 4.8 \times 10^{-8}\,\% = 0.012$
 water vapour has greatest greenhouse effect [2]

(b) (i) new sea areas so more marine organisms or named organism/formerly cold area grows new warm-climate plants [1]

 (ii) deserts formed which reduce food availability/cold-adapted organisms not suited to new climate/terrestrial organisms destroyed by the rising seas [1]

[Total: 4]

2

(a) no significant migration;
 no significant births or deaths;
 marking does not have an adverse effect. [3]

(b) S = total number of individuals in the total population
 S_1 = number captured in sample one, marked and released, i.e. 16
 S_2 = total number captured in sample two, i.e. 12
 S_3 = total marked individuals captured in sample two, i.e. 5

 $$\frac{S}{S_1} = \frac{S_2}{S_3} \quad \text{so, } S = \frac{S_1 \times S_2}{S_3}$$

 $S = \dfrac{16 \times 12}{5}$ Estimated no. of shrews is 38.4 [2]

(c) Not very reliable because the numbers are quite low. High population numbers are more reliable. [1]

[Total: 6]

Chapter 6 Disease, exercise and health

1

(a) droplet infection [1]

(b) (i) carriers [1]

    (ii) we do not know that they carry the pathogen as they display no symptoms, so the people do not avoid contact and pass on the bacteria [2]

(c) (i) antibiotics or named antibiotics [1]

    (ii) Phagocyte – engulfs/produces pseudopodia/ phagocytosis; digests the bacterium/causes lysis of the bacterium [2]

B-lymphocyte – changes into plasma cell; makes antibodies [2]

T-lymphocyte – whole cell links to bacterial antigen sites; cell is usually destroyed by this; reacts to bacterial antigen [2]

[Total: 11]

2

(a) (i) ability to secrete more antibodies [1]

    (ii) the antibodies have specific receptor sites which bind with the antigens; they have a flexible protein which changes angle to fit the antigens; antibodies result in the destruction of the antigen in some way/ neutralise toxin/cluster around antigens then cause precipitation/cause agglutination [3]

    (iii) are produced when body first exposed to antigen; remain in body to react quickly when exposed to same antigen again [2]

(b) when the immune system attacks the person's own cells; pernicious anaemia/rheumatoid arthritis [2]

(c) the influenza virus often mutates so lymphocytes take longer to produce antibodies [1]

[Total: 9]

Notes

Notes

Index